T0280821

Functional Metal-Oxide Nanostructures

MATERIALS RESEARCH SOCIETY
SYMPOSIUM PROCEEDINGS VOLUME 1174

Functional Metal-Oxide Nanostructures

Symposium held April 14–17, San Francisco, California, U.S.A.

EDITORS:

Junqiao Wu
University of California, Berkeley
Berkeley, California, U.S.A.

Wei-Qiang Han
Brookhaven National Laboratory
Upton, New York, U.S.A.

Anderson Janotti
University of California, Santa Barbara
Santa Barbara, California, U.S.A.

Ho-Cheol Kim
IBM Almaden Research Center
San Jose, California, U.S.A.

Materials Research Society
Warrendale, Pennsylvania

CAMBRIDGE UNIVERSITY PRESS
Cambridge, New York, Melbourne, Madrid, Cape Town,
Singapore, São Paulo, Delhi, Mexico City

Cambridge University Press
32 Avenue of the Americas, New York NY 10013-2473, USA

Published in the United States of America by Cambridge University Press, New York

www.cambridge.org
Information on this title: www.cambridge.org/9781107408241

Materials Research Society
506 Keystone Drive, Warrendale, PA 15086
http://www.mrs.org

© Materials Research Society 2009

First published 2009
First paperback edition 2012

Single article reprints from this publication are available through
University Microfilms Inc., 300 North Zeeb Road, Ann Arbor, MI 48106

CODEN: MRSPDH

ISBN 978-1-107-40824-1 Paperback

SYNTHESIS

*Invited Paper

PROPERTIES

APPLICATIONS

PREFACE

These are the proceedings of Symposium V, "Functional Metal-Oxide Nanostructures," held April 14–17 at the 2009 MRS Spring Meeting in San Francisco, California.

Metal oxides and their nanostructures have emerged as an important class of materials with a rich spectrum of properties and great potential for device applications. These include transparent electrodes, high electron mobility transistors, gas sensors, photovoltaic and photonic devices, energy harvesting devices with multiferroic oxides, non-volatile memories with defective oxides, etc. The symposium provided a unique opportunity for the materials community to discuss the fundamental problems related to synthesis, characterization, assembly, and device applications of oxide nanostructures, as well as the most recent progress in these fields. Presentations in the symposium covered a wide range of topics such as physics and applications of phase transitions, strain and interfaces in complex oxide nanostructures, nano-ionics and resistance switching in oxides, assembly and processing of oxide nanostructures, zinc oxide, titanium dioxide, other oxide semiconductors and their applications.

The symposium attracted a large group of attendees from around the world, reflecting the great current excitement in this field. The symposium consisted of nine oral sessions and two poster sessions, with 14 invited talks, 120 contributed talks, and over 50 posters. For some of the invited talks, the large meeting room was fully crowded with over a hundred attendees. One of our posters won the best poster award.

We would like to acknowledge the funding support from IBM Almaden Research Center, JSR Micro Inc., Samsung Cheil Industries, Inc., and Samsung Electronics Co., Ltd.

Junqiao Wu
Wei-Qiang Han
Anderson Janotti
Ho-Cheol Kim

June 2009

MATERIALS RESEARCH SOCIETY SYMPOSIUM PROCEEDINGS

MATERIALS RESEARCH SOCIETY SYMPOSIUM PROCEEDINGS

Prior Materials Research Society Symposium Proceedings available by contacting Materials Research Society

Synthesis

Mater. Res. Soc. Symp. Proc. Vol. 1174 © 2009 Materials Research Society 1174-V09-31

Fabrication of Nano-Sized Tin Oxide Powder by Spray Pyrolysis Process

Jaekeun Yu, Jwayeon Kim and Jeoungsu Han

Department of Advanced Materials Engineering, Hoseo University, Asan 336-795, Korea

ABSTRACT

By using tin chloride solution as the raw material, a nano-sized tin oxide powder with average particle size below 50 nm is generated by spray pyrolysis reaction. This study also examines the influences of the reaction parameters such as reaction temperature and the concentration of raw material solution on the powder properties. As the reaction temperature increases from 800 to 850 ℃, the average particle size of the generated powder increases from 20 nm to 30 nm. As the reaction temperature reaches 900 ℃, the droplets are composed of nano-particles with average size of 30 nm, while the average size of individual particles increases remarkably up to 80~100 nm. When the tin concentration reaches 75 g/L, the average particle size of the powder is below 20 nm. When the tin concentration reaches 150 g/L, the droplets are composed of nano particles with average size around 30 nm, whereas the average size of independent particles increases up to 80~100 nm. When the concentration reaches 400 g/L, the droplets are composed of nano-particles with average size of 30 nm.

INTRODUCTION

Spray pyrolysis reaction[1-7] is the method of manufacturing nano-sized metal oxide powder. In this reaction, chemical components are uniformly blended in the solution state so as to make a complex solution, which is in turn sprayed into a reaction furnace with a high temperature. In the furnace, spray pyrolysis reaction is accomplished instantly, and as a result, the ultra-fine metal oxide powder is formed. The purpose of this study is to develop a technology for the mass production of nano-sized tin oxide powder with uniform particle size and average particle size below 50 nm in a spray pyrolysis reaction device by using tin chloride solution as the raw material. This study also examines the influences of reaction parameters on the properties of the generated tin oxide powder. These parameters include the reaction temperature and the concentration of raw material solution.

EXPERIMENT

In this study, a tin oxide powder with average particle size below 50 nm is manufactured from the tin chloride solution with the presence of tetravalent tin ions by using spray pyrolysis process. First, tin powder with purity above 99.9 % is dissolved into the hydrochloric acid

solution with concentration of 25 % until the concentration of tin constituent is adjusted to 400 g/L. A certain amount of ammonia is added in the solution so as to oxidize all the divalent tin ions into tetravalent ones. Then, the solution is filtered triply through filter paper, and the purified solution is utilized as the raw material of the spray pyrolysis process. The constituents of silicon dioxide, phosphorus, calcium, chromium and copper in this solution are below 100 ppm. This solution is diluted by distilled water so that the tin concentration of the solution is respectively adjusted to 400, 300, 150 and 75 g/L.

In order to generate the nano-sized tin oxide powder, a spray pyrolysis system is specially designed and built for this study. The schematic diagram of this system is shown in Figure 1. The properties of the generated powder are examined according to TEM analysis (the generation of single crystal particles), SEM analysis (particle size distribution and particle shape), and XRD analysis (powder phase and composition).

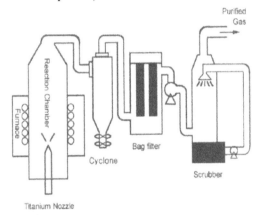

Figure 1. Schematic diagram of spray pyrolysis system

DISCUSSION

The Powder properties influenced by the reaction temperature

Figure 2 is the 30,000 times magnified image under SEM, which shows the property change of the generated powder relative to the rise of reaction temperature from 800 ℃ to 950 ℃. Figure 3 is the 200,000 times magnified image under SEM, which shows the property change of the generated powder under the same reaction conditions as of Figure 2.

When the reaction temperature is at 800 ℃ as shown in Figure 2 (a), the generated powder maintains the spherical shape of droplets micronized at the initial stage of pyrolysis reaction, the particle size is larger than those at the other reaction temperatures, and the particle size distribution appears to be comparatively uniform. According to the result shown in Figure 3 (a), it can be concluded that the droplets shown in Figure 2 (a) are composed of nano-sized particles with average particle size around 20 nm. When the reaction temperature increases up to 900 ℃,

4

there are extremely severe bursts of droplets at the initial stage of pyrolysis reaction. According to the result shown in Figure 3 (c), the droplets shown in Figure 2 (c) consist of nano-sized particles with average particle size around 30 nm, while the particles are combined much more compactly in contrast to the case at 850 ℃. For the independent particles, however, the average particle size remarkably increases up to 80~100 nm, and the particle surface becomes comparatively compact. When the reaction temperature increases up to 950 ℃, there is a much severer burst of droplets at the initial stage of pyrolysis reaction in contrast to the case at 900 ℃. As shown in Figure 2 (d), among all the powder generated by pyrolysis reaction, the ratio of the powder that maintains the spherical shape of droplets micronized at the initial stage decreases sharp, and the majority of the generated powder are in the shape of independent particles.

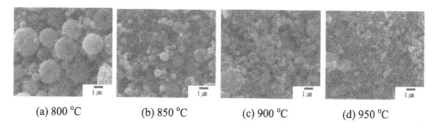

    (a) 800 °C   (b) 850 °C   (c) 900 °C   (d) 950 °C

Figure 2. SEM photographs of produced powder according to various reaction temperatures at raw material solution of 150 g/ℓ Sn, 20 mL/min. inlet speed of solution, 2 mm nozzle tip size and 3 kg/cm^2 air pressure. (X 30,000)

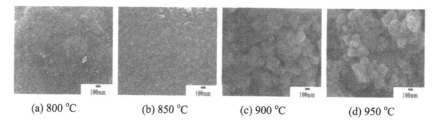

    (a) 800 °C   (b) 850 °C   (c) 900 °C   (d) 950 °C

Figure 3. SEM photographs of produced powder according to various reaction temperatures at raw material solution of 150 g/ℓ Sn, 20 mL/min. inlet speed of solution, 2 mm nozzle tip size and 3 kg/cm^2 air pressure. (X 200,000)

Figure 4 is the result of TEM analysis, (a) which shows the structural property of the generated powder, and (b)the diffraction pattern of one particle for identifying the single crystal of oxide particles. Because other particles have the similar patterns, the compact single crystal structure of the generated powder can be confirmed by this analysis.

(a) TEM photographs of produced powder.   (b) Selective diffraction pattern of single particle.

Figure 4. TEM photographs of produced powder and selective diffraction pattern(zone axis $[\bar{1}10]$) of single particle at reaction temperature of 900℃, raw material solution of 150g/L tin, 20 mL/min. inlet speed of solution, 2 mm nozzle tip size and 3 kg/cm² air pressure.

Fig. 5 is the results of XRD analysis under the same reaction conditions as shown in Fig. 2. As the reaction temperature increases from 800 to 900℃, the intensities of the first, second and third peaks greatly increases, resulting the combined effect of the following facts: a) the average particle size of independent single crystals increases remarkably; and b) the single crystals inside the particles, which are in the shape of droplet, are much compactly combined together. When the reaction temperature increases ups to 950℃, the overall peak intensity decreases greatly, resulting the fact: at the temperature of 950℃, the ratio of the droplet types, in which tin oxide single crystals aggregate, decreases sharply; and the average particle size of single crystals is smaller in contrast to the case at 900℃.

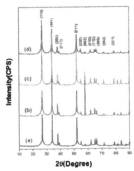

Figure 5. XRD patterns of produced powder according to reaction temperature at raw material solution of 150g/ℓ Sn, 20 mL/min. inlet speed of solution, 2 mm nozzle tip size and 3 kg/cm² air pressure.   (a) 800°C   (b) 850°C   (c) 900°C   (d) 950°C

The powder properties influenced by concentration of the raw material solution

Figure 6 and Figure 7 are the 30,000 and 200,000 times magnified images under SEM shown the property change of the generated powder relative to the concentration of raw material solution rising from 75 g/L to 400 g/L, respectively. When the tin concentration of raw material solution is at the lowest 75 g/L, there is no significant difference of concentration between the surface and the center of droplet till the evaporation of solvent inside droplet ends completely. As a result, there is no severe burst of droplets during the pyrolysis reaction process. As shown in Figure 6 (a), there are two coexisting shapes among the powder generated by pyrolysis reaction such as droplets micronized at the initial stage of pyrolysis reaction and independent particles. According to Figure 7 (a), the droplets shown in Figure 6 (a) are composed of extremely small particles with average particle size below 20 nm. When the tin concentration increases up to 150 g/L, the droplet size increases remarkably contrast to the case at 75 g/L. The droplets eventually burst and the particle size distribution becomes more irregular. As shown in Figure 6 (b), the generated powder maintains the spherical shape of droplets micronized at the initial stage of pyrolysis reaction in spite of the severe burst of droplets, and the average particle size increases greatly. Moreover, the ratio of droplet-shaped particles to independent particles decreases remarkably.

| (a) 75 g/L | (b) 150 g/L | (c) 300 g/L | (d) 400 g/L |

Figure 6. SEM photographs of produced powder according to various tin concentration of raw material solution at 900℃, 20 mL/min. inlet speed of solution, 2 mm nozzle tip size and 3 kg/cm^2 air pressure. (X 30,000)

| (a) 75 g/L | (b) 150 g/L | (c) 300 g/L | (d) 400 g/L |

Figure 7. SEM photographs of produced powder according to various tin concentration of raw material solution at 900℃, 20 mL/min. inlet speed of solution, 2 mm nozzle tip size and 3 kg/cm^2 air pressure. (X 200,000)

According to Figure 7 (b), the droplets shown in Figure 6 (b) consist of nano-sized particles with average particle size around 30 nm, and are rather more compactly combined together in contrast to the case at 75 g/L. For the case of independent particles, the average particle size increases to a significant level of 80~100 nm, and the particle surface becomes comparatively compact. When the concentration increases up to 400 g/L, which is close to the saturated concentration. Because the solute is excessively saturated on the droplet surface at the initial stage of pyrolysis reaction, the decrease of droplet size caused by the evaporation of solvent is almost undetectable, and the droplet size increases greatly in contrast to the cases at low concentrations. However, because of the severe droplet burst during the pyrolysis process, the particle size distribution appears extremely irregular and the particle surface also appears in the extremely inhomogeneous shape. Also, the evaporation of solvent completes extremely fast and the evaporation heat of solvent decreases greatly during the pyrolysis process. As a result, the sintering reaction proceeds further and the droplets combine mutually each others. According to Figure 7 (d), the droplets shown in Figure 6 (d) consist of nano-sized particles with average particle size around 30 nm, and are rather compactly combined together. This result is due to the combined effect of the following two mechanisms: a) the droplet size increases with the rise of concentration after the evaporation of solvent; and b) the burst of droplets becomes severer with the rise of concentration during the pyrolysis reaction.

CONCLUSIONS

As the reaction temperature increases from 800 to 850 ℃, the average particle size of the generated powder increases from 20 nm to 30 nm, and most of the powder is in the shape of droplet. As the reaction temperature reaches 900 ℃, the droplet-shaped particles are composed of nano-particles with average size of 30 nm, while the average size of individual particles increases remarkably up to 80~100 nm. As the reaction temperature reaches 950 ℃, most of the particles appear mutually independent, while the average particle size is around 80 nm.

For the raw material solution with concentration of 75 g/L, the average particle size of the generated powder is below 20 nm, and there is a strong tendency of cohesion between particles. When the tin concentration reaches 150 g/L, the droplets are composed of nano particles with average size around 30 nm, whereas the average size of independent particles increases remarkably up to 80~100 nm. When the concentration reaches 400 g/L, the droplets are composed of nano-particles with average size of 30nm, and most of the droplets are compactly combined together.

ACKNOWLEDGMENTS

This work was supported by the Regional Innovation Center of Hoseo University in Korea.

REFERENCES

1. J.K. Yu, S.G. Kang, K.C. Chung. J.S. Han and D.H. Kim, Mater. Trans. **48**, 249 (2007).
2. J.K. Yu, S.G. Kang, J.B. Kim, J.Y. Kim, J.S. Han, J.W. Yoo, S.W. Lee and Z.S.Ahn, Mater. Trans. **47,** 1695 (2006).
3. J.K. Yu, G.H Kim, T.S. Kim and J.Y. Kim, Mater. Trans. **46,** 1695 (2005).
4. T.G. Carreno, M.P. Morales and C.J. Serna, Mater. Lett. **43,** 97 (2000),
5. D. Majumdar, T.A. Shefelbine and T.T. Kodas, J. Mater. Res. **11,** 2861 (1996).
6. M.A.A. Elmasry, A. Gaber and E.M.H. Khater, Powder Technology **90,** 165 (1997).
7. S.C. Zhang and G.L. Messing, J. Am. Ceram. Soc. **73,** 61 (1990).

Mater. Res. Soc. Symp. Proc. Vol. 1174 © 2009 Materials Research Society 1174-V03-06

Synthesis and Characterization of Nanocarbon-Supported Titanium Dioxide

Marcus A. Worsley, Joshua D. Kuntz, Octavio Cervantes, T. Yong-Jin Han, Peter J. Pauzauskie, Joe H. Satcher, Jr. and Theodore F. Baumann
Physical and Life Sciences Directorate, Lawrence Livermore National Laboratory, 7000 East Avenue, Livermore, CA 94550, U.S.A.

ABSTRACT

In this report, we describe recent efforts in fabricating new nanocarbon-supported titanium dioxide structures that exhibit high surface area and improved electrical conductivity. Nanocarbons consisting of single-walled carbon nanotubes and carbon aerogel nanoparticles were used to support titanium dioxide particles and produce monoliths with densities as low as 80 mg/cm^3. The electrical conductivity of the nanocarbon-supported titanium dioxide was dictated by the conductivity of the nanocarbon support while the pore structure was dominated by the titanium dioxide aerogel particles. The conductivity of the monoliths presented here was 72 S/m and the surface area was 203 m^2/g.

INTRODUCTION

Titanium dioxide is a widely researched material with applications ranging from photocatalysts to electrodes to hydrogen storage materials [1-9]. However, issues such as absorption limited to the ultraviolet range, high rates of electron-hole recombination, and relatively low surface areas have limited commercial use of titanium dioxide. Recent efforts have focused on combining titanium dioxide with various materials to address some of these issues [8-24]. Titanium dioxide in the presence of carbon (e.g. carbon nanotubes (CNT)) is currently one of the most attractive combinations [13-25]. While recent work has shown some improvements, surfaces areas and photocatalytic activity are still limited. Maintaining high surface areas while improving electrical conductivities, one could envision charging-discharging rates and photoefficiencies that are significantly higher than currently possible. Unfortunately for CNT composites, improvements in electrical conductivity are often not fully realized due to poor dispersion of CNTs in the matrix material, impeding the formation of a conductive network. However, with a mechanically robust, electrically conductive CNT foam, one could imagine simply coating this low-density CNT scaffold with titanium dioxide, yielding conductive nanocarbon-supported titanium dioxide.

Here we present the synthesis and characterization of such a high-surface area, conductive TiO$_2$/CNT composite. We recently reported the synthesis of a novel CNT-based foam, consisting of bundles of single-walled nanotubes (SWNT) crosslinked by carbon aerogel (CA) nanoparticles, which would serve as an excellent candidate for the CNT scaffold of the TiO$_2$/CNT composite. The SWNT-CA foams simultaneously exhibited increased stiffness, and high electrical conductivity even at densities approaching 10 mg cm^{-3} without reinforcement [25]. The foams are stable to temperatures approaching 1000°C and have been shown to be unaltered by exposure to extremely low temperatures during immersion in cryogenic liquids [26]. So, in addition to their use in applications such as catalyst supports, sensors, and electrodes, these ultralight, robust foams could allow the formation of novel CNT composites. As the conductive network is already established, it can be impregnated through the wicking

9

process [27] with a matrix of choice, ranging from inorganic sols to polymer melts to ceramic pastes. Thus, a variety of conductive CNT composites could be created using the SWNT-CA foam as a pre-made CNT scaffold. In this report, we use the SWNT-CA as a scaffold for the synthesis of conductive, high surface area TiO_2/CNT composites.

EXPERIMENT

Materials. All reagents were used without further purification. Resorcinol (99%) and formaldehyde (37% in water) were purchased from Aldrich Chemical Co. Sodium carbonate (anhydrous) was purchased from J.T. Baker Chemical Co. Highly purified SWNTs were purchased from Carbon Solutions, Inc.

SWNT-CA preparation. The SWNT-CAs were prepared as described in previous work [25]. Briefly, in a typical reaction, purified SWNTs (Carbon Solutions, Inc.) were suspended in deionized water and thoroughly dispersed using a VWR Scientific Model 75T Aquasonic (sonic power ~ 90 W, frequency ~ 40 kHz). The concentration of SWNTs in the reaction mixture was 0.7 wt%. Once the SWNTs were dispersed, resorcinol (1.235 g, 11.2 mmol), formaldehyde (1.791 g, 22.1 mmol) and sodium carbonate catalyst (5.95 mg, 0.056 mmol) were added to the reaction solution. The resorcinol to catalyst ratios (R/C) employed was 200. The amount of resorcinol and formaldehyde (RF solids) used was 4 wt%. The sol-gel mixture was then transferred to glass molds, sealed and cured in an oven at 85°C for 72 h. The resulting gels were then removed from the molds and washed with acetone for 72 h to remove all the water from the pores of the gel network. The wet gels were subsequently dried with supercritical CO_2 and pyrolyzed at 1050°C under a N_2 atmosphere for 3 h. The SWNT-CAs materials were isolated as black cylindrical monoliths. Foams with SWNT loadings of 30 wt% (0.5 vol%) were prepared by this method.

TiO_2/SWNT-CA composite preparation. Sol-gel chemistry was used to deposit the TiO_2 aerogel layer on the inner surfaces of the SWNT-CA support. The TiO_2 sol-gel solution was prepared as described in previous work [28]. In a typical synthesis, SWNT-CA parts were immersed in the TiO_2 sol-gel solution and full infiltration of the SWNT-CA pore network by the sol-gel solution was achieved under vacuum. Following gelation of the titania network, the wet composite was dried using supercritical CO_2, yielding the TiO_2/SWNT-CA composite.

Characterization. Bulk densities of the TiO_2/SWNT-CA composites were determined from the physical dimensions and mass of each sample. The volume percent of SWNT in each sample was calculated from the initial mass of SWNTs added, assuming a CNT density of 1.3 g/cm^3, and the final volume of the aerogel. Scanning electron microscopy (SEM) characterization was performed on a JEOL 7401-F at 10 keV (20mA) in SEI mode with a working distance of 2 mm. Transmission electron microscopy (TEM) characterization was performed on a JEOL JEM-200CX. Thermogravimetric analysis (TGA) was performed on a Shimadzu TGA 50 Thermogravimetric Analyzer to determine TiO_2 content. Samples were heated in flowing air at 10 sccm to 1000°C at 10°C/min in alumina boats. The weight fraction of material remaining was assumed to be pure stoichiometric TiO_2. Energy dispersive spectroscopy confirmed that only TiO_2 remained after TGA was performed. Surface area determination and pore volume and size analysis were performed by Brunauer-Emmett-Teller (BET) and Barrett-Joyner-Halenda (BJH) methods using an ASAP 2000 Surface Area Analyzer (Micromeritics Instrument Corporation) [29]. Samples of approximately 0.1 g were heated to 300°C under vacuum (10^{-5} Torr) for at least 24 hours to remove all adsorbed species. Electrical conductivity was measured using the four-probe method similar to previous studies [25, 30, 31]. Metal electrodes were attached to the ends of the cylindrical samples. The amount of current transmitted through the sample during measurement was 100 mA and the voltage drop along the sample was measured over distances of 3 to 6 mm. Seven or more measurements were taken on each sample.

DISCUSSION

The microstructure of the TiO_2/SWNT-CA composites was examined using SEM and TEM. (Figure 1) As shown in Figure 1, the network structure of the TiO_2/SWNT-CA composites is similar to that observed in pristine SWNT-CA. The presence of the TiO_2 aerogel layer on the surface of the nanotube bundles can be seen in TEM image. Interestingly, the TiO_2 aerogel appears to have formed primarily on the surfaces of the nanotube bundles despite the fact that the TiO_2 sol-gel solution filled the entire pore volume of the support. The open pore volume in the TiO_2/SWNT-CA composite is only sparsely populated with TiO_2 particles. This observation indicates that nucleation of the TiO_2 particles during the sol-gel reaction preferentially occurs at the surface of the nanotube bundles.

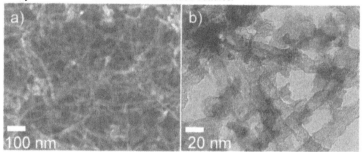

Figure 1. a) SEM and b) TEM images of TiO_2/SWNT-CA.

Thermal gravimetric analysis in air was used to determine the TiO_2 content in the as-TiO_2/SWNT-CA composites. (Figure 2) As expected, combustion of the pristine SWNT-CA occurs around 500°C and the material is completely consumed by 600°C. The 5 wt% remaining is likely metal catalyst from the CNTs. The titania exhibits an initial mass loss generally attributed to moisture and organics below 300°C and is stable thereafter. Not surprisingly, the TGA plot for TiO_2/SWNT-CA material is a composite of the plots for titania and the SWNT-CA. It is interesting to note that the combustion of the SWNT-CA occurs significantly earlier for the TiO_2/SWNT-CA compared to that for the pristine SWNT-CA, which may be the result of a catalytic effect of the titania aerogel particles on carbon oxidation. Nevertheless, the nearly 50 wt% remaining after combustion of the SWNT-CA confirm the presence of titania in the TiO_2/SWNT-CA composite.

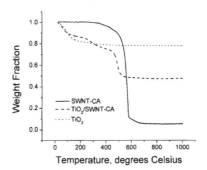

Figure 2. TGA plot of SWNT-CA, TiO₂/SWNT-CA, and TiO_2 in air.

Figure 3 plots the pore size distribution of the SWNT-CA, TiO₂/SWNT-CA composite, and pristine TiO_2 aerogel. The BET surface area, electrical conductivity and other physical properties of these materials are summarized in Table I. Table I shows that the TiO₂/SWNT-CA composite has high surface area and electrical conductivity. In fact, the electrical conductivity of the SWNT-CA is not adversely affected by the infiltration of the insulating material. Though, based on the SEM and TEM images (Figure 1), the titania aerogel appears to simply coat the SWNT-CA scaffold, the increased surface area suggests that the pore morphology of the titania dominates the overall pore morphology of the composite. This is confirmed via the pore size distribution, which shows that the pore size distribution of the TiO₂/SWNT-CA is much closer to that of pristine TiO_2 aerogel than that of the SWNT-CA. Thus, with the TiO₂/SWNT-CA composite, a new class of materials with good electrical conductivity and high surface area are realized.

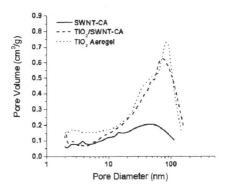

Figure 3. Semi-log plot of the pore size distribution of the SWNT-CA, TiO₂/SWNT-CA, and TiO_2 aerogel.

Table I. Physical properties of SWNT-CA, TiO$_2$/SWNT-CA, and TiO$_2$ aerogel.

Material	CNT, vol% (wt%)	Density, g/cm^3	S_{BET}, m^2/g	σ, Scm^{-1}
SWNT-CA	0.5 (30)	0.030	184	0.77
TiO$_2$/SWNT-CA	0.5 (8)	0.082	203	0.72
TiO$_2$ aerogel	0 (0)	0.193	237	<0.001

CONCLUSIONS

In summary, we have described a straightforward method for the fabrication of electrically conductive, high-surface area TiO$_2$/CNT composites. The novel TiO$_2$/SWNT-CA monoliths was prepared by coating the CNT struts within the SWNT-CA scaffold with amorphous sol-gel-derived TiO$_2$ particles. Given the technological interest in crystalline TiO$_2$, work is in progress to convert the amorphous TiO$_2$ layer to the anatase crystalline phase. The conductive network of the SWNT-CA scaffold remained intact after infiltration yielding a composite with a conductivity of 72 S m^{-1} and a surface area of 203 m^2 g^{-1}. Therefore, the SWNT-CAs were shown to provide the means to create conductive, high-surface area TiO$_2$ composites. The general nature of this method should provide a route for the synthesis of a variety of conductive, high surface area composites with applications in photocatalysis and energy storage.

ACKNOWLEDGMENTS

This work was performed under the auspices of the U.S. Department of Energy by Lawrence Livermore National Laboratory under Contract DE-AC52-07NA27344 and funded by the DOE Office of Energy Efficiency and Renewable Energy.

REFERENCES

[1] XB Yu, DA Grant, GS Walker *Journal of Physical Chemistry C* **2008**, *112*, 11059.
[2] A Fischer, P Makowski, JO Mueller, M Antonietti, A Thomas, F Goettmann *Chemsuschem* **2008**, *1*, 444.
[3] YZ Li, NH Lee, DS Hwang, JS Song, EG Lee, SJ Kim *Langmuir* **2004**, *20*, 10838.
[4] H Yamada, T Yamato, R Hidaka, I Moriguchi, T Kudo *Solid State Ionics: the Science and Technology of Ions in Motion* **2004**, 533.
[5] M Gratzel *Current Opinion in Colloid & Interface Science* **1999**, *4*, 314.
[6] M Schneider, A Baiker *Catalysis Today* **1997**, *35*, 339.
[7] LK Campbell, BK Na, EI Ko *Chemistry of Materials* **1992**, *4*, 1329.
[8] YF Zhu, L Zhang, L Wang, Y Fu, LL Cao *Journal of Materials Chemistry* **2001**, *11*, 1864.
[9] J Retuert, R Quijada, VM Fuenzalida *Journal of Materials Chemistry* **2000**, *10*, 2818.

[10] KGK Warrier, SR Kumar, CP Sibu, G Werner *Journal of Porous Materials* **2001**, *8*, 311.

[11] Z Ding, X Hu, PL Yue, GQ Lu, PF Greenfield *Catalysis Today* **2001**, *68*, 173.

[12] T Torimoto, S Ito, S Kuwabata, H Yoneyama *Environmental Science & Technology* **1996**, *30*, 1275.

[13] B Liu, HC Zeng *Chemistry of Materials* **2008**, *20*, 2711.

[14] HT Yu, X Quan, S Chen, HM Zhao *Journal of Physical Chemistry C* **2007**, *111*, 12987.

[15] WD Wang, CG Silva, JL Faria *Applied Catalysis B-Environmental* **2007**, *70*, 470.

[16] HS Shin, YS Jang, Y Lee, Y Jung, SB Kim, HC Choi *Advanced Materials* **2007**, *19*, 2873.

[17] GM An, WH Ma, ZY Sun, ZM Liu, BX Han, SD Miao, ZJ Miao, KL Ding *Carbon* **2007**, *45*, 1795.

[18] XB Yan, BK Tay, Y Yang *Journal of Physical Chemistry B* **2006**, *110*, 25844.

[19] I Moriguchi, R Hidaka, H Yamada, T Kudo, H Murakami, N Nakashima *Advanced Materials* **2006**, *18*, 69.

[20] Y Yu, JC Yu, JG Yu, YC Kwok, YK Che, JC Zhao, L Ding, WK Ge, PK Wong *Applied Catalysis a-General* **2005**, *289*, 186.

[21] YZ Li, DS Hwang, NH Lee, SJ Kim *Chemical Physics Letters* **2005**, *404*, 25.

[22] L Han, W Wu, FL Kirk, J Luo, MM Maye, NN Kariuki, YH Lin, CM Wang, CJ Zhong *Langmuir* **2004**, *20*, 6019.

[23] B Tryba, AW Morawski, M Inagaki *Applied Catalysis B: Environmental* **2003**, *46*, 203.

[24] S Sakthivel, H Kisch *Angewandte Chemie-International Edition* **2003**, *42*, 4908.

[25] MA Worsley, SO Kucheyev, JH Satcher, AV Hamza, TF Baumann *Applied Physics Letters* **2009**, *94*, 073115.

[26] T Wiley, personal communication.

[27] J Wang, L Angnes, H Tobias, RA Roesner, KC Hong, RS Glass, FM Kong, RW Pekala *Analytical Chemistry* **1993**, *65*, 2300.

[28] SO Kucheyev, TF Baumann, YM Wang, T van Buuren, JH Satcher *Journal of Electron Spectroscopy and Related Phenomena* **2005**, *144*, 609.

[29] SJ Gregg, KSW Sing, *Adsorption, Surface Area and Porosity* (Academic, London, ed. 2nd, 1982), pp.

[30] XP Lu, O Nilsson, J Fricke, RW Pekala *Journal of Applied Physics* **1993**, *73*, 581.

[31] MA Worsley, JH Satcher, TF Baumann *Langmuir* **2008**, *24*, 9763.

Mater. Res. Soc. Symp. Proc. Vol. 1174 © 2009 Materials Research Society   1174-V09-32

A Single-Step Route Towards Large-Scale Deposition of Nanocomposite Thin Films Using Preformed Gold Nanoparticles

Weiliang Wang[1], Kevin Cassar[2], Steve Sheard[1], Peter Dobson[1], Simon Hurst[3], Peter Bishop[2], Ivan Parkin[4]

[1]Engineering Science, Oxford University, Oxford, United Kingdom
[2]Johnson Matthey Technology Centre, Reading, United Kingdom
[3]Pilkington Technology Centre, Lathom, United Kingdom
[4]Chemistry, University College London, London, United Kingdom

ABSTRACT

We have been developing noble metal nanoparticles and nanocomposites for large scale application to glass surfaces. The novel functionality of the nanocomposites is attributed to the properties of both the metal nanoparticles and host matrix. Here a single-process route to nanocomposite thin films by spray deposition technique has been investigated. Preformed gold nanoparticles have been incorporated into several different transition metal oxides (TiO_2, SnO_2, ZnO). The nanocomposite films showed intense coloration due to the surface plasmon resonance effects of gold nanoparticles embedded in the host matrix. The gold nanoparticles were found well embedded into the host metal oxides homogeneously. This film deposition method can easily be scaled up and is compatible with current industrial on-line processes.

INTRODUCTION

Nanocomposite materials based on noble nanoparticles (NPs) within a metal oxide matrix have been an intensive area of research due to their intriguing functional properties that can be tailored by the nanoparticle size, shape and surrounding medium [1-3]. Noble metal nanoparticles, e.g. Au, Ag, Cu, generally exhibit a strong absorption peak in the visible range of the spectrum, due to the surface plasmon resonance (SPR) effect, which has led to application in optical devices [4, 5]. Dielectric materials, especially transition metal oxides, have been extensively investigated in their own right due to their functional properties, such as photocatalysis, electrochromism and solar control [6-9]. Nanocomposites comprising a host material and embedded metal particles therefore bear the combined functionalities of both phases and can be carefully selected to exploit the desired properties of either material. Such kind of nanocomposites have shown potential applications in glass manufacturing, catalysis and electronics industries [10].

Here we report a single-step process for production of nanocomposite thin films using a preformed gold colloid. Several different transition metal oxides (TiO_2, SnO_2, ZnO) are investigated as the host materials. The deposition is based on the spray pyrolysis technique which is simple, flexible and capable of scale up for large area deposition with potentially a low processing cost. We have successfully deposited the composite thin films with Au nanoparticles homogeneously embedded in the host matrix.

EXPERIMENT

Gold nanoparticles were synthesised and dispersed in water or ethanol, in which the resultant concentration of gold is 0.14%. The precursors for TiO_2, SnO_2 and ZnO are supplied from Johnson Matthey Catalyst and used as received. One-pot dispersions for deposition of composite films were prepared by mixing gold colloids and different precursors for TiO_2, SnO_2 or ZnO (with a concentration of 0.1 M) in deionised water or ethanol. The molar ratio of gold to the transition metal varied from 1:10 to 1:20.

The set-up for the laboratory-scale spray deposition has been described elsewhere [11]. Typically, the precursor solutions were injected by the syringe pump under a constant flow rate (1-2 ml/min) and spread out by the compressed air at a pressure of 20-30 psi (138-206 kPa). The atomized droplets were sprayed onto Pilkington float glass substrates (4mm thick, with a SiO_2 barrier layer on top). Substrates were heated to the range of 150-550 °C. The coated glass samples received further heat treatment in most cases by placing them in a furnace at a temperature in the range of 400-600 °C for 30-60 minutes to remove any remaining organic ligands and further enhance the decomposition of the precursor.

UV-vis spectra were obtained using a Varian Cary 5000 UV-visible-NIR spectrometer. Thickness measurements were carried out using a Veeco Dektak 6M stylus profilometer. In order to obtain information about the morphology of the films and nanoparticle structure, transmission electron microscopy (JEOL 4000HR) and scanning electron microscopy (JSM 6500) micrographs were taken.

DISCUSSION

Preformed Au nanoparticles and Au films

The initial gold colloids were transparent and a deep red color. TEM imaging of the colloid on a holey carbon film showed the gold nanoparticles to have a mean diameter of 20 nm with a narrow size distribution (Figure 1). Gold nanoparticle thin films were spray deposited onto glass at 180 °C using this solution. The films were red in appearance when viewed in transmitted light and an intense SPR absorption peak at 542 nm was displayed in the UV-visible spectrum.

Au/TiO₂ thin films using preformed Au nanoparticles.

The one-pot solution containing gold colloids and the titanium precursor in water was used for spray deposition of Au/TiO₂ nanocomposite thin films. The mixture solution again shows a clear deep red color (just from the surface plasmon resonance of gold nanoparticles) and the stability of the solution was maintained for over two weeks. This solution was atomized and sprayed at 200 °C on Pilkington float glass. The as-sprayed composite films appeared red to the transmitted light. UV-vis spectroscopy of the film showed an absorption band in the region of 543 nm indicating the deposition of gold nanoparticles in the film. Annealing treatment was applied at 400 °C or 600 °C for 30 minutes respectively. With annealing the color of the film changed from red to purple (400 °C) and to blue (600 °C). Optical spectroscopy showed an absorption peak shift from 543 nm to 572 nm and to 599 nm, respectively (Figure 2). The film thickness decreased from 728 nm to 320 nm and 240 nm. This can be explained by the effects of annealing as removing organic ligands and causing densification of the TiO_2 host, which results in a higher refractive index of the host material and shorter neighboring nanoparticle distance [12, 13].

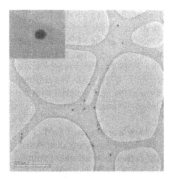

Figure 1. TEM images of gold nanoparticle colloid on a holey carbon film with a mean diameter of 20 nm. The scale bar in the micrograph is 0.5 μm.

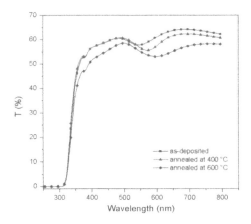

Figure 2. Optical absorbance spectra of gold nanoparticle/TiO_2 host thin film on Pilkington float glass before and after annealing.

In order to compare the "spray followed by annealing" process, a direct spray deposition without annealing was conducted using the same solution but at a range of different deposition temperatures from 200 °C to 550 °C. Transparent thin films with colouration were found to be successfully deposited at all these temperatures. These films were robust under washing with good adherence to the glass substrate. The gold nanoparticles could not be removed from the TiO_2 host by repeated washing, indicating the gold nanoparticles are embedded in the film. It was also found that the deposition temperature has an influence on the optical properties of the composite thin films. The colour of the films deposited showed a transition from red to blue in transmitted light with increasing deposition temperature. UV-vis spectroscopy showed that at relatively low deposition temperatures (200-400 °C), significant red-shift of the SPR absorption peak was observed with increasing substrate temperature; estimated peaks occurred at 544, 562 and 585 nm with deposition temperatures of 200, 300 and 400 °C. While for temperatures above 400 °C, less red-shift was observed with peak around 600 nm at 550 °C (Figure 3). We attribute the SPR red-shift to the formation and densification of TiO_2 anatase in the host material with the increasing substrate temperature, increasing the dielectric constant of the host material.

SEM imaging conducted in a secondary electron mode shows the presence of gold nanoparticles at or just below the surface of the film. The particles appear randomly distributed and around 20 nm in diameter. Backscattering mode shows bright particles on a darker texture background, which is interpreted as gold nanoparticles embedded in a titania host matrix due to a higher atomic weight of gold. EDAX also confirms the presence of gold and titanium.

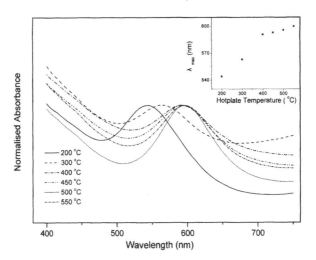

Figure 3. Measured normalized absorbance spectra of gold nanoparticles/TiO_2 thin films deposited with substrate temperatures from 200 °C to 550 °C which shows a shift of the SPR peak with increasing substrate temperature (see inset). The peaks show a red-shift from 545 nm to 600 nm.

Au/SnO₂ and Au/ZnO composite films using preformed Au nanoparticles

To extend the knowledge of this process to other host metal oxides, single-step deposition of Au/SnO$_2$ and Au/ZnO composite films using preformed Au nanoparticles were carried out in a similar manner of depositing Au/TiO$_2$ as described previously. These films deposited were found to be robust with good adherence to the glass surface and the gold particles were well embedded in the host materials. XRD confirmed the presence of Au, SnO$_2$ and ZnO in the films. The films were transparent showing intense blue appearance in transmission. Figure 4 shows the UV-vis spectra of three composite films comprising gold nanoparticles in different host materials of TiO$_2$, SnO$_2$ and ZnO. It was observed that the SPR absorption bands occur at different positions for Au/TiO$_2$ (602 nm), Au/SnO$_2$ (595 nm) and Au/ZnO (580 nm). The red-shift of the SPR band for Au/TiO$_2$ compared to Au/SnO$_2$ and Au/ZnO is due to the higher dielectric constants of TiO$_2$ in that $\varepsilon(TiO_2) > \varepsilon(SnO_2) > \varepsilon(ZnO)$ [3].

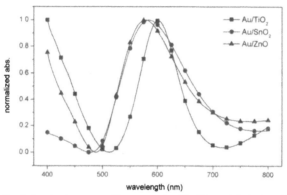

Figure 4. Normalized absorbance spectra for composite thin films with Au NPs in different host materials of TiO$_2$, SnO$_2$, ZnO.

The incorporation of gold nanoparticles in the host matrix

In the composite thin film where the host material serves to protect and stabilize the particles, the method employed to embed the particles in the host matrix is an important process to be considered. In the single-step deposition via a chemical route, successful deposition requires the preparation of a stable and homogeneous suspension containing the precursors of both the particles and host materials. There are constraints imposed on obtaining a stable single-source suspension due to issues relating to pH, choice of solvents and concentration which might result in an immiscible solution and subsequent flocculation. Therefore careful investigations have been made in our experiments to determine compatible precursor components for preparation of the one-pot suspension. Also the incorporation of gold nanoparticles into the host material during the deposition process is essential for successful deposition of homogeneous composite thin films. The gold nanoparticles might be prevented from embedding into the host material during the deposition due to the action of thermophoresis [1]. In our spray deposition experiments, however, it is assured that the driving force from the compressed air ejected from the nozzle can overcome the repelling thermophoretic force.

CONCLUSIONS

It has been demonstrated that composite films can be spray deposited by incorporating preformed Au NPs into different host materials of TiO_2, SnO_2 and ZnO. A single step process using a one-pot solution has been sprayed onto glass substrates at different substrate temperatures. The optical properties of the composite films can be adjusted by varying the properties of metal nanoparticles, host material precursors and substrate temperature. Spray deposition offers a simple, rapid and low cost approach to large area coatings for glass building products. Deposition using preformed nanoparticles maintains the size distribution of the NP and leads to more predictable coloration of the resulting films. This film deposition method can be scaled up and is compatible with current industrial on-line processes.

ACKNOWLEDGMENTS

This work has been funded by the UK Technology Strategy Board. We thank Guillermo Benito, Troy Manning and Paolo Melgari for helpful discussion as well as Professor Patrick Grant of Materials Department, Oxford University for providing access to the spray equipment.

REFERENCES

1. R. Palgrave, I. P. Parkin, J. Am. Chem. Soc. **128**,1587–1597 (2006).
2. X. Xu , M. Stevens, M. B. Cortie, Chem. Mater. **16**, 2259 (2004).
3. D. Buso, J. Pacifico, A. Martucci and P. Mulvaney Adv. Funct. Mater. **17**, 347 (2007).
4. L. Armelao, D. Barreca, G. Bottaro, A. Gasparotto, S. Gross, C. Maragno, E. Tondello, Coord. Chem. Rev. **250**, 1294 (2006).
5. T. Ung, L. M. Liz-Marzan, P. Mulvaney, Colloids and Surfaces.A-Physicochem. Eng. Aspects **202**, 119 (2002).
6. D. Lahiri, V. Subramanian, T. Shibata, E. E. Wolf, B. A. Bunker, P. V. Kamat, J. Appl. Phys. **93**, 2575 (2003).
7. S. Ashraf, C. S. Blackman, G. Hyett, I. P. Parkin, J. Mater. Chem. **16**, 3575 (2006).
8. T. He, Y. Ma, Y. Cao, W. Yang, J. Yao Phys. Chem. Chem. Phys. **4**, 1637 (2002).
9. S. K. Medda, S. De and G. De, J. Mater. Chem. **15**, 3278–3284 (2005).
10. G. Walters and I. P. Parkin, J. Mater. Chem. **19**, 574 (2009)
11. X. Zhao, C. Hinchliffe, C. Johnston, P. J. Dobson and P. S. Grant, Mater. Sci. Eng. B, **151** 140 (2008).
12. S. Deki, Y. Aoi, H. Yanagimoto, K. Ishii, K. Akamatsu, M. Mizuhata and A. Kajinami, J. Mater. Chem, 6, 1879 (1996).
13. M. Lee. L. Chae, and K. C. Lee, Nanostructured Materials **11**, 195 (1999).

Mater. Res. Soc. Symp. Proc. Vol. 1174 © 2009 Materials Research Society   1174-V11-05

Ordered Nanoparticle Arrays Synthesized From Self-Assembled Diblock Copolymer Templates

Qiang Fu, Anita Ghia, Chi-shuo Chen, Jennifer Lu
School of Engineering, University of California, Merced
Merced CA 95348

ABSTRACT

We present a comprehensive study of using diblock copolymer micelle templates to synthesize ordered nanoparticle arrays. Ionic and coordination bonds have been exploited to incorporate nanoparticle precursors into cores of block copolymer micelles. Polystyrene-b-poly (4-vinylpyridine) (PS-b-P4VP) has been shown to be able to localize anions via electrostatic attraction with protonated pyridine cations while transitional metals can be sequestered through coordination bonds. Polystyrene-b-poly (acrylic acid) (PS-PAA) can localize a variety of cations via ionic bonds with acrylic anions. We have demonstrated that the size of nanoparticles can be tuned by controlling the solution concentration of an ionic precursor. By mixing these two distinct block copolymers which can selectively interact with different precursor species, complex nanoparticle architectures can be generated thus paving a path for new applications.

1. INTRODUCTION

There have been burgeoning research efforts aiming for establishing controllable nanoparticle synthesis due to their technological potential in catalysis, sensing, medicine and electronic and optoelectronics [1-4]. Many of these applications require nanoparticles with controllable size, compositions, spacing and location. In the past decade, biological templates and synthetic templates [5-6] have been explored to generate nanoscale morphologies. Among them, templating self-assembled block copolymers has emerged as one of the most promising approaches for nanoparticle formation because of its simplicity, versatility and low cost.

Diblock copolymers are polymers with two different polymer blocks connected by a single covalent bond. In selective surroundings such as a solvent preferential for one block, diblock copolymers will phase separate into various morphologies [7]. For example, hexagonal micelles formed by self-assembly of diblock copolymers in solution followed by coating into monolayer films have been used to localize nanoparticle precursors to form ordered nanoparticle arrays on substrates [8]. The size and spacing of nanoparticles can be controlled by adjusting the block length and loading ratio of metal precursors [9-10]. In addition, diblock copolymers are compatible with standard semiconductor fabrication processes and can be readily implemented in microelectronic and optical device fabrication.

Essential to the employment of diblock copolymer templates for nanoparticle formation is the ability to sequester nanoparticle precursors selectively onto one of the blocks. Herein we report, by exploiting electrostatic and coordination interactions between nanoparticle precursors and functional groups of the repeat units of a diblock copolymer, a variety of new and existing types of uniform nanoparticles have been successfully generated. By utilizing two different types of micelles which have different chemical and physical affinity, hybrid and complex nanoparticle arrays can be created.

2. EXPERIMENTAL

2.1 Incorporating precursors in solution micelles

Polystyrene (42100)–b-poly (4-vinyl pyridine) (8100) (0.25 wt%) (PS-P4VP1, Polymer Source Inc, Quebec, Canada) was dissolved in toluene by stirring for 2 hrs. HAuCl$_4$ and Cu(OAc)$_2$ (Sigma-Aldrich, Missouri, USA) were then added into the polymer solution and the solution was left to react overnight. The polymer solutions were then spin-coated on silicon substrates at 3000 rpm. Finally the polymer templates were then removed by UV-Ozone treatment at 100°C for 20 mins to form the nanoparticle array.

2.2 Incorporating precursors in surface micelles

Toluene solutions of polystyrene (109000)-b-poly (4-vinyl pyridine) (27000) (PS-P4VP2) and polystyrene (16500)-b-poly (acrylic acid) (4500) (PS-PAA) were deposited on silicon substrates by spin-coating at 3000 rpm. The polymer films were then baked at 100°C for 5mins. To introduce the precursor ions, the substrates were immersed in either a 1mM HAuCl$_4$ or a 10μM ZnCl$_2$ aqueous solution for 2 mins. The substrates were then dried with N$_2$ and treated with UV Ozone to generate the nanoparticle arrays.

2.3 Selective incorporation of precursors in surface micelles formed by two distinct micelles

A solution containing both PS-P4VP2 and PS-PAA (molar ratio 0.25:1) was spin-coated on the substrates. The substrates were then immersed in either 1mM HAuCl$_4$ or 1uM CoCl$_2$ aqueous solution for 2 mins respectively. After dried with N$_2$, the substrates were treated with UV Ozone at 100°C for 20 mins to remove the polymer template.

3. RESULTS and DISCUSSION

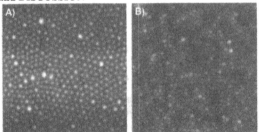

Figure 1. A) An AFM image of PS-P4VP1 micelles deposited on substrates. (Height scale: 25 nm). B) An AFM image of PS-PAA micelles deposited on substrates. (Height scale: 10 nm, Scan size: 1 μm)

3.1 Polymer micelle formation

Two different diblock copolymers, PS-b-P4VP and PS-b-PAA, were used in this investigation. When dissolved in toluene, both polymers form spherical micelles. Hydrophilic blocks, P4VP and PAA, form the cores of the micelles while the PS blocks form the corona of the micelles. The micelles are stable and can be transferred to substrates by simply spin-coating without disaggregation. Figure 1 shows a set of AFM images showing micelle thin films prepared on silicon substrates. The density and size difference between these two result primarily from difference in block length.

3.2 Incorporating precursors in solution micelles

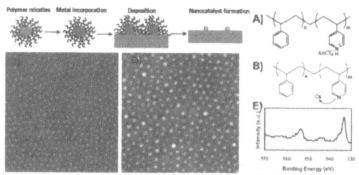

Figure 2. An illustration showing incorporation of precursors into micelles in solution for nanoparticle arrays. A) Incorporation of HAuCl4 to PS-b-P4VP1 by electrostatic attraction. B) Incorporating Cu(OAc)2 to PS-PAA by complex formation. C) AFM image of gold nanoparticle arrays with PS-P4VP1 as templates. D) CuO nanoparticle arrays with PS-b-P4VP1 as templates. (Scan size: 1 μm, height scale: 10 nm)

HAuCl4 and Cu(OAc)2 were chosen as precursors in this investigation to study the ability of PS-b-P4VP to localize nanoparticle precursors in solution micelles. Figure 2A is the schematic process flow of creating nanoparticles by incorporating inorganic species into solution micelles. We have found that both HAuCl4 and Cu(OAc)2 can be effectively incorporated into PS-P4VP micelles. HAuCl4 binds to the pyridine group by electrostatic attraction between protonated pyridine units and AuCl4⁻ while Cu species bind to pyridine via complex formation as illustrated in Figure 2A. After removing the polymer templates by UV Ozone treatment, ordered Au and CuO nanoparticle arrays have been successfully generated (Figure 2C, D). Figure 2E is the XPS analysis result showing that resulting nanoparticles are indeed Cu oxide.

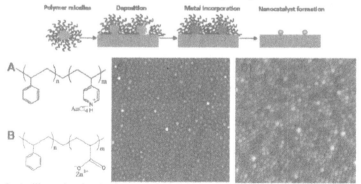

Figure 3. An illustration showing incorporation of precursors into micelles on surface for nanoparticle arrays. A) Incorporation of HAuCl4 to PS-b-P4VP1 by electrostatic attraction. B) Incorporating Zn²⁺ to PS-b-PAA by electrostatic attraction. C) AFM image of gold nanoparticle arrays with PS-b-P4VP1 as templates. D) ZnO nanoparticle arrays with PS-b-PAA as templates. (Scan size: 1 μm, height scale: 10 nm)

3.2 Incorporating precursors in surface micelles

3.2.1 Nanoparticle formation

Efficient incorporation of the precursors into solution micelles requires precursors to be soluble in the solvent used for the formation of block copolymer micelles, hydrophobic toluene in this case. Often many precursor compounds are hydrophilic. To overcome this problem, surface sequestration has been investigated. Figure 3A illustrates the nanoparticle formation by exposing surface micelles to a precursor solution. Similarly, the interaction between functional groups in a block and precursors is employed to selectively localize precursors into the cores of the block copolymer micelles as depicted in Figure 3B. Figure 3C and 3D are AFM images of Au nanoparticles formed with PS-P4VP1 templates, in which $HAuCl_4$ has been sequestered into PS-P4VP1 surface micelles. After removal of polymer templates, Au nanoparticle arrays have been formed. Correspondingly, cationic precursors, Zn^{2+}, can be incorporated into PS-PAA micelles, in which acrylic anion can effectively bind to zinc cations via electrostatic attraction. The ZnO nanoparticles prepared by PS-b-PAA have higher density and narrower spacing than the Au nanoparticles made from PS-b-P4VP1 as template, resembling their parent surface micelle topography shown by the AFM images in Figure 1.

3.2.2 Dynamic process of incorporation of precursors into surface micelles

The size of nanoparticles is determined by the aggregation number of a polymer micelle and precursor loading. In solution, it is well known that nanoparticle size can be controlled by adjusting the amount of precursor species introduced. In thin films, we have found that the size of the nanoparticles can be modulated by controlling the equilibrium between free ions in solution and available binding sites available in the core of surface micelles via adjusting precursor concentration. Figure 4 shows nanoparticle arrays formed on a surface by immersing the micelle film in solutions with different $HAuCl_4$ concentration. The density and spacing of the nanoparticles are similar but the size increases with higher concentration of $HAuCl_4$, indicating more $AuCl_4^-$ ion binding to P4VP cores. When the $HAuCl_4$ concentration is above 2 mM, the

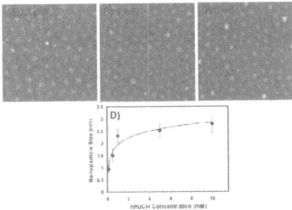

Figure 4. (A-C) AFM images of nanoparticle arrays by incorporating $HAuCl_4$ into PS-b-P4VP1 from aqueous solutions with different concentrations. D) A graph exhibiting the correlation between Au nanoparticle size and $HAuCl_4$ concentration. (Scan size: 400 nm, height scale: 10 nm).

nanoparticle size became constant implies that saturation has reached no more P4VP binding sites available.

3.3 Selective incorporation of precursors in surface micelles formed by two distinct micelles

Having established the methods of creating diverse nanoparticles via solution and surface micelles, single or binary nanoparticles arranged in a complex architecture becomes possible. By mixing PS-b-PAA with PS-b-P4VP2 solution micelles first and then depositing them onto a surface, a unique morphology consisting of PS-b-P4VP and PS-b-PAA surface micelles can be formed on a surface as shown in Figure 5A. In addition, this morphology can be modified by adjusting the volume ratio of these two block copolymers. The findings will be reported in the near future. The larger micelles in Figure 5A are formed by PS-b-P4VP2 and smaller micelles are produced by PS-b-PAA. When exposed to a HAuCl₄ solution, the anionic AuCl⁻₄ can only be incorporated in PS-b-P4VP2 micelles due to favorable electrostatic attraction between AuCl⁻₄ and positively charged protonated pyridine units. Therefore Au species can be selectively incorporated into the PS-b-P4VP2 micelles. Contrarily, due to the repulsive interaction between AuCl⁻₄ and negatively charged acrylic anions, Au nanoparticles can't be formed in the areas where PS-b-PAA micelles are located after the removal of the polymer template as shown in Figure 5B. However, if the micelle film is immersed in a $CoCl_2$ solution, in which Co^{2+} can bind to PAA block via electrostatic attraction and P4VP by complexation, nanoparticle arrays resembling the surface order created by their template, ordered PS-P4VP2 and PS-PAA surface micelles can be generated, as can be seen in Figure 5C. Such a versatile and simple approach could be useful to prepare nanoparticle arrays with multiple types of nanoparticles on the same surface. Potential applications range from catalysis, sensing to serving as active components in optoelectronic devices.

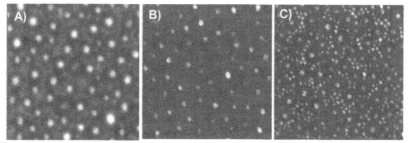

Figure 5. A) An AFM image of micelle films with both PS-P4VP2 and PS-PAA micelles on surface. (Height scale: 45 nm). B) Gold nanoparticle arrays formed on surface by immersing the micelles in 10mM HAuCl₄ solution. (Height scale, 20 nm) C) CoO_x nanoparticles formed on surface by immersion the micelle films in 5mM $CoCl_2$ solution. (Height scale: 15 nm) (Scan size: 1 μm)

4. CONCULSIONS

In summary, we report a systematic study on employing self-assembled block copolymers as templates for creating ordered and uniform sized nanoparticle arrays. Both anionic and cationic metal species can be efficiently incorporated into the block copolymer templates for the synthesis of nanoparticle arrays. In addition, complex formation can be used to sequester transition metal species. Ordered Au, ZnO, CoO_x and CuO nanoparticle arrays have been generated via either the solution or surface approach. We have demonstrated a facile approach to control nanoparticle size with the surface micelle approach through adjusting reaction equilibrium between nanoparticle precursors in solution and reaction sites in the micelle cores. Furthermore, we have done exploratory work to show that, by employing selective interaction between precursors and block copolymers, precursor species can be incorporated into one or two different types of micelle templates on a surface. Diverse nanoparticles arranged in simple or complex architecture thus can be generated. It is expected that future exploration on this front will lead to the expansion of on-going research activities and pave new scientific and technological innovative paths in the field nanotechnology.

REFERENCES
1.   M.E. Akerman, W.C.W. Chan, P. Laakkonen, S.N. Bhatia, and E. Ruoslahti, *Proceedings of the National Academy of Sciences of the United States of America* 99, 12617 (2002).
2.   P. Alivisatos, *Nature Biotechnology* 22, 47 (2004).
3.   A.N. Shipway, E. Katz, and I. Willner, *Chemphyschem* 1, 18 (2000).
4.   S.H. Sun, C.B. Murray, D. Weller, L. Folks, and A. Moser, *Science* 287, 1989 (2000).
5.   A.K. Boal, F. Ilhan, J.E. DeRouchey, T. Thurn-Albrecht, T.P. Russell, and V.M. Rotello, *Nature* 404, 746 (2000).
6.   C.A. Mirkin, R.L. Letsinger, R.C. Mucic, and J.J. Storhoff, *Nature* 382, 607 (1996).
7.   F.S. Bates, and G.H. Fredrickson, *Annual Review of Physical Chemistry* 41, 525 (1990).
8.   J.P. Spatz, S. Mossmer, C. Hartmann, M. Moller, T. Herzog, M. Krieger, H.G. Boyen, P. Ziemann, and B. Kabius, *Langmuir* 16, 407 (2000).
9.   J.P. Spatz, A. Roescher, and M. Moller, *Advanced Materials* 8, 337 (1996).
10.   J. Lu, S.S. Yi, T. Kopley, C. Qian, J. Liu, and E. Gulari, *Journal of Physical Chemistry B* 110, 6655 (2006).

Mater. Res. Soc. Symp. Proc. Vol. 1174 © 2009 Materials Research Society   1174-V04-12

Strain Relaxation of Self-Nanostructured Solution Derived La$_{0.7}$Sr$_{0.3}$MnO$_3$ Films

P. Abellan[1], F. Sandiumenge[1], C. Moreno[1], M. J. Casanove[2], T. Puig[1] and X. Obradors[1]
[1]Institut de Ciència de Materials de Barcelona, CSIC, 08193 Bellaterra, Catalonia, Spain
[2]CEMES, 29 rue Jeanne Marvig, BP 94347, 31055 Toulouse Cedex 4, France

ABSTRACT

The morphological and microstructural evolution associated with an exsolution driven self-nanostructuration process of La$_{0.7}$Sr$_{0.3}$MnO$_3$ films, is investigated using scanning force microscopy, reciprocal space mapping and transmission electron microscopy. The focus is placed on the misfit strain relaxation mechanism. Surfaces with atomically flat terraces are already developed after 1hour at 1000 °C while first fingerprints of phase exsolution do not appear until 9-10 hours. X-ray diffraction reciprocal-space mapping reveals that 24 nm thick films remain strained during the whole microstructural evolution, while 12 hour annealed films undergo almost total plastic relaxation of the misfit strain at a thickness of 60 nm. Overall, these results point to a kinetic limitation of dislocation mechanisms. It is argued that chemical relaxation provides a significant contribution to misfit strain relief.

INTRODUCTION

Manganese perovskites exhibit a rich electronic and magnetic behavior including colossal magnetoresistance, large spin polarization, electronic phase segregation, orbital ordering and charge ordering [1], which have generated great expectations in diverse technological fields. A marked singularity of such materials is that they exhibit strong correlations between several degrees of freedom (lattice, charge, spin, orbital) that are in turn strongly sensitive to external perturbations, among which misfit strain outstands as the most suffered one. Taking advantage of property-microstructure correlations obviously involves a deep understanding of microstructural mechanisms to relief misfit strains. In this context, a main conclusion can be inferred from studies on manganite epitaxial films [2-5], namely, that these phases remain (apparently) strained up to thicknesses that may involve elastic energies even one order of magnitude larger than those required for the formation of a misfit dislocation array assuming a classical surface nucleation dislocation loop mechanism [6]. This anomalous behavior in fact unveils an extremely complex lattice response to misfit strain, presumably involving a subtle interplay between chemical disorder and coupled orbital-charge-lattice effects [7].

We have recently demonstrated that solution derived La$_{0.7}$Sr$_{0.3}$MnO$_3$ (LSMO) films submitted to extended thermal annealing periods, undergo an unprecedented spontaneous self-nanostructuration process [8]. The process is thermodynamically driven by the exsolution of (La,Sr)O$_x$ islands and Sr$_3$Mn$_2$O$_7$ type Ruddlesden-Popper (RP) inclusions, which self assemble within a room temperature ferromagnetic (T$_c$=360 K , M$_s$=590 emu/cm^3) film of high crystalline perfection [9]. The study of this system is interesting because it constitutes a single-step self-nanostructuration mechanism with exciting implications on surface patterning of a metallic ferromagnetic film with isolating islands. In the present work we focus on the evolution of the strain state of the LSMO matrix during the microstructural development leading to the formation of the nanocomposite consisting of (Sr, La)O$_x$ islands outcropping the film surface and inclusions of the Ruddlesden-Popper (RP) $Sr_3Mn_2O_7$ phase epitaxially embedded in a LSMO film exhibiting room temperature ferromagnetic behaviour [8].

EXPERIMENTAL

$La_{1-x}Sr_xMnO_3$ (x=0.3) thin films have been prepared by metal-organic decomposition (MOD) [8]. The precursors were propionates containing La, Sr and Mn with nominal stoichiometric ratio 0.7:0.3:1.0. Precursors were dissolved in propionic acid solvent, filtered and deposited on a $SrTiO_3$(001) (hereafter STO) substrate by spin coating. The error in the solution metal content was found to be within ~5 at.% by Induced Coupled Plasma (ICP) analyses. As-deposited precursor films were subsequently annealed for 1 hour at 900 °C and 1000 °C, and for 6 hours, 9 hours and 12 hours at 1000 °C in order to track the evolution of the surface and microstructural features. Films prepared using solution concentrations (defined with respect to the manganese content) of 0.3 M, 0.4 M and 0.5 M, leading to thicknesses of 24 nm, 35 nm and 60 nm, respectively, were also prepared in order to investigate the thickness dependence of the misfit strain relaxation.

The surface topography of the films was systematically characterized by scanning force microscopy (SFM). The average strain of the films was determined by X-ray diffraction reciprocal space mapping. Transmission electron microscopy (TEM) investigations were performed using the 200 kV JEM-2011 and 200 kV FEI C_s corrected Tecnai F20 electron microscopes. Cross-sectional and planar view TEM specimens were prepared by using the mechanical tripod polisher technique followed by Ar milling in a Precision Ion Polishing System (PIPS).

RESULTS

Development of the film nanostructure

TEM observations reveal that after 1 hour annealing at 900 °C the films are already re-crystallized into an epitaxial film without any evidence of phase exsolution. SFM images reveal that at 1000 °C the surface is flat at all annealing stages until first signatures of island outcropping appear after 9.

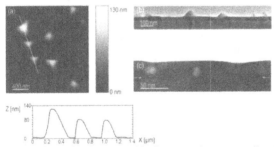

Figure 1. (a) SFM topographic image and corresponding line scan showing several facetted triangular pyramidal islands. (b)-(c) are low magnification TEM images where $(La,Sr)O_x$ islands (b), as well as the RP inclusions embedded in the LSMO film (c), are observed.

28

After 12 hours annealing at 1000 °C, the surface features well facetted triangular pyramidal islands with heights comprised between about 40 nm and 130 nm, outcropping on the atomically flat LSMO film. **Figure 1(a)** shows an SFM topographic image along with its corresponding line profile illustrating the topographic features of these films. As discussed above, the spontaneous outcropping of $(Sr,La)O_x$ pyramidal islands, also shown in the [100] cross sectional TEM image presented in **figure 1(b)**, occurs simultaneously with the formation of rounded inclusions of the Ruddlesden-Popper (RP) $Sr_3Mn_2O_7$ phase [8], evidenced by a lighter contrast in **figure 1(c)**. Notably, the magnetic properties of the LSMO matrix, in particular a ferromagnetic transition at T_c=360 K and a saturation magnetization of M_s=590 emu/cm^3 [8], bare witness of a precise overall stoichiometric balance between the islands and the RP inclusions.

Misfit strain relaxation behavior during microstructural evolution

The average strain of the films was investigated by X-ray reciprocal space mapping. Results show that the 24 nm films are strained at all processing stages. In agreement with this, cross sectional high resolution TEM revealed almost dislocation free interfaces. A Poisson ratio of υ=0.38 was derived from X-ray and electron diffraction, in agreement with values derived by other authors, υ=0.34 [2], and υ=0.35 [9] for similar non nanostructured PLD films. As the film becomes thicker, its in-plane and out-of-plane lattice parameters get closer to the one in the bulk compound (**figure 2**).

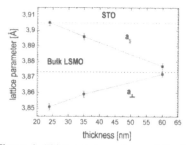

Figure 2. Thickness dependence of the in-plane ($a_{//}$) and off-plane ($a_\perp$) lattice parameters, as determined from reciprocal space mapping around the (103) film and substrate reflections for 24 nm, 35 nm and 60 nm films annealed for 12 hours, with fully developed nanostructure.

In agreement with the volume averaged view given by the above X-ray diffraction study, both cross-section high resolution, and diffraction contrast planar view TEM observations of the 24 nm film revealed an almost dislocation free interface. Misfit dislocations occur, however in some areas. Similarly, {100} type twins appear occasionally. **Figures 3(a)** and **(b)** are planar view bright field (BF) **g**=010 images of two different regions containing misfit dislocations and {100} twins normal to the film. Their orientation perpendicular to the interface can be inferred from the absence of so-called δ-fringes arising from a difference in the excitation parameters in the two twin variants. Polygonal shape $(Sr,La)O_x$ islands are marked in (**a**) by white doted contours. These areas are exclusively composed by a single twinning variant, along the [100] direction. **Figures 3(c)** and **(d)**, show examples of BF images taken with **g** =110 and **g** =100 in an untwined area. With **g** =100 **(d)** the dislocation set with **b** perpendicular to **g** is out of contrast.

DISCUSSION

According to the **g·b** criterion, the contrast behavior of the dislocations is consistent with b=<100> edge dislocations. Theoretically, the misfit dislocation spacing for full relaxation is given by S=|b|/ε ≈51nm, where the value |b|=3.89Å has been used. Our results clearly demonstrate that the misfit dislocation content of the interface lies far below that value. As far as the twinning is considered, its irregular distribution hardly allows thinking that they can contribute efficiently to misfit relief. Indeed, synchrotron radiation diffraction studies reveal that the twin angle vanishes towards the interface, indicating that the first unit cells ahead of the interface are clamped to the cubic substrate [10]. Note that the twin rotation associated with {100} twins perpendicular to the film is about an axis parallel to the interface, which would even reduce the projected matching distance on the substrate plane, thus further increasing the tensile misfit strain on STO. Therefore, it is very unlikely that such twins contribute to the relaxation of misfit tensile strains in LSMO films.

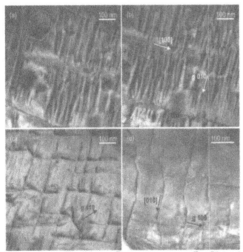

Figure 3. Two-beam bright field TEM images taken with diffraction vectors as indicated in the images. (a), (b) correspond to twinned areas, and (c) and (d) correspond to untwined areas.

Considering the introduction of half loops as the operating relaxation mechanism, and using isotropic elasticity theory, the equilibrium thickness h_c for misfit strain relaxation is given by [6]:

$$h_c = (Kbd_f)/(4\pi M\varepsilon_s d_s) \ln((\beta h_c)/(b))$$

where ε is the elastic accommodation strain, $M=C_{11}+C_{12}-2(C^2_{12}C_{11})$ the biaxial modulus, **b** the Burgers vector and β is the cut-off radius of the dislocation core, generally comprised between 1 and 4. This yields h_c values comprised between ≈ 7 nm ($\beta=1$) and ≈ 12 nm ($\beta=4$), well below the thickness of the present film, ~ 24 nm. Surprisingly, however, 12 hours processed films are not observed to relax up to a thickness of 60 nm (**figure 2**). Since the elastic energy of the strained film linearly scales with its thickness [6], the observed relaxation thickness is too far from h_c to be considered without the interplay of relaxation mechanisms other than pure dislocation ones. Despite its magnitude, the present critical thickness is even notably smaller than that reported for similar PLD films, >100 nm [2], suggesting the role of additional relaxation mechanisms.

As far as the influence of the growth mechanism is concerned, MOD growth is characterized by very high undercoolings which favor the massive nucleation of LSMO grains within the precursor volume in a single nucleation step [11,12]. At this precise stage, misfit stresses affecting the first layer of crystallites epitaxially nucleated on the substrates, are likely to be relieved by a dislocation wall mechanism [13], and this might explain the fact that MOD derived films tend to relax at a very small thickness [14]. However, subsequent epitaxial growth proceeds through a secondary growth mechanism from the substrate to the film surface. As the slab of transformed precursor thickens, misfit stresses building up within the growing film can in principle be relieved by dislocation emission from the advancing grain boundaries towards the LSMO/STO interface. However, this is not observed. This apparent difficulty for dislocations to expand out of the grain boundaries may be caused by large Peierls-Nabarro stresses expected for LSMO. This kinetic limitation for dislocation relaxation has been recognized by other authors in PLD films [2,4].

Manganese perovskites, however, may admit high levels of chemical disorder, either in the form of cation concentration fluctuations or as point defect associations. Indeed, an enrichment in Sr along the growth direction, towards the surface of the film, has been reported in the literature, [2,15] which according to the anticorrelation between Sr content and lattice parameter [16], could assist for a partial relief of the misfit strain, though this effect is insufficient for a complete relaxation at the LSMO/STO interface. Point defect associations, relevant to ionic crystals, can also respond to mechanical stress through a shift of their association/dissociation equilibrium [17]. Particularly relevant to misfit epitaxy, is the conclusion that mechanical stress can be partially relieved by the change in linear dimensions of the lattice induced by this mechanism, as demonstrated in films of the ionic conductor $Ce_{0.8}Gd_{0.2}O_{1.9}$ [17].

Dislocation mechanisms appearing inefficient from the first growth stages, the most obvious differential feature that could explain the contrast between the relaxation thickness of the present films and PLD ones [2], is the nanostructure development. The fact that in 24 nm films it only develops after long periods of thermal annealing (9 hours at 1000°C) without signatures on the average strain state of the films (as determined by reciprocal space mapping), indicates that the kinetics of the transformation is rather insensitive to mechanical stresses, i.e., its development takes place once the film has presumably adopted a stable configuration, playing therefore a marginal role. It is possible, however, that the complex interplay between phase instability, chemical and mechanical stresses is modified in thicker films, in a way that plastic relaxation is favored at a thickness smaller than in the absence of the nanostructure.

CONCLUDING REMARKS

Self-nanostructured LSMO films consisting of $(La,Sr)O_x$ islands and RP inclusions within an epitaxial matrix of high crystalline perfection, are shown to plastically relax at a thickness of 60nm, about six times larger than the equilibrium critical thickness predicted for systems exhibiting pure dislocation mechanisms. The development of the nanostructure, takes place only after 9-10 hours of annealing at 1000 °C without any signature on the strain state of the films, thus indicating that the driving force for phase exsolution is essentially thermodynamic. It is argued that the actual critical thickness in the present system is determined by a subtle balance between elastic energy and the chemical energy associated with chemical disorder before the nanostructure is developed.

ACKNOWLEDGMENTS

We acknowledge the financial support from spanish MEC (NANOSELECT, CSD2007-00041 FPU, AP2005-4669 FPU, MAT2008.01022/NAN), Generalitat de Catalunya (Catalan Pla de Recerca SGR-0029 and CeRMAE), CSIC (PIF-CANNAMUS). The Cs corrected F20-FEI electron microscope was used through the European project ESTEEM (Contract number 026019). We also acknowledge electron microscopy facilities from Servei de Microscòpia de la UAB.

REFERENCES

1. Y. Tokura, Rep. Prog. Phys. 69 (2006) 797.
2. J.L. Maurice, F. Pailloux, A. Barthélémy, O. Durand, D. Imhoff, R. Lyonnett, A. Rocher, J.-P. Contour, Phil. Mag. Vol. 83, No. 28 (2003) 3201
3. E. Gommert, H. Cerva, A. V. Rucki, R. Helmholt, J. Wecker, C. Kuhrt, K. Samwer, J. Appl. Phys. 81 (1997) 5496.
4. O. I. Lebedev, G. Van Tendeloo, S. Amelinckx, H. L. Hu, K. M. Krishnan, Phil. Mag. A 80 (2000) 673.
5. M. –J. Casanove, C. Roucau, P. Baulès, J. –C. Ousset, D. Magnoux, J. –F. Bobo, Appl. Surf. Sci. 188 (2002) 19.
6. J. W. Matthews, J. Vac. Sci. Technol. 12 (1975) 126.
7. Ll. Abad, V. Laukhin, S. Valencia, A. Gaup, W. Gudat, Ll. Balcells, B. Martínez, Adv. Funct. Mater. 17 (2007) 3918.
8. C. Moreno, P. Abellan, A. Hassini, A. Ruyter, A. Perez del Pino, F. Sandiumenge, M-J. Casanove, J. Santiso, T. Puig, X. Obradors, Adv. Funct. Mater., 2009, 10.1002/adfm.200900095
9. L. Ranno, A. Llobet, R. Tiron, E. Favre-Nicolin, Appl. Surf. Sci. 188 (2002) 170.
10. U. Gebhardt, N. V. Kasper, A. Vigliante, P. Wochner, H. Dosch, F. S. Razavi, H.-U. Habermeier, Phys. Rev. Lett. 98 (2007) 096101.
11. R. W. Schwartz, Chem. Mater. 9 (1997) 2325
12. A. Cavallaro, F. Sandiumenge, J. Gàzquez, T. Puig, X. Obradors, J. Arbiol, H. C. Freyhardt, Adv. Funct. Mater. 16 (2006) 1363
13. I.A. Ovid'Ko, J. Phys. Condens. Matter 11 (1999) 6521
14. P.A. Langjahr, F.F. Lange, T. Wagner, M. Rühle, Acta Mater. 46 (1998) 773
15. T. T. Fister, D. Dillon, D. Fong, J. A. Eastman, P. M. Baldo, M. J. Highland, P. H. Fuoss, K. R. Balasubramaniam, J. C. Meador, P. A. Salvador, Appl. Phys. Lett. 93 (2008) 151904.
16. A. Hammouche, E. Siebert, A. Hammou, Mater. Res. Bull. 24 (1989) 367.
17. M. Greenberg, E. Wachtel, I. Lubomirsky, J. Fleig, J. Maier, Adv. Funct. Mater. 16 (2006) 48; A. Kossoy, Y. Feldman, E. Wachtel, I. Lubomirsky, J. Maier, Adv. Funct. Mater. 17 (2007) 2393.

Mater. Res. Soc. Symp. Proc. Vol. 1174 © 2009 Materials Research Society   1174-V11-08

Quantitative Investigation of the Factors Affecting the Hydrothermal Growth of Zinc Oxide Nanowires

Aron R. Rachamim[1], Sharvari H. Dalal[1], Sieglinde M.-L. Pfaendler[1], Michael E. Swanwick[1], Andrew J. Flewitt[1] and William I. Milne[1]

[1] Electrical Engineering Division, University of Cambridge, 9 JJ Thomson Avenue, Cambridge, CB3 0FA, U.K.

ABSTRACT

Zinc oxide (ZnO) nanowires (NWs) are receiving significant industrial and academic attention for a variety of novel electronic, optoelectronic and MEMS device applications due to their unusual combination of physical properties, including being optically transparent, semiconducting and piezoelectric. Hydrothermal growth is possible at significantly lower temperatures (and hence lower thermal budgets) compared with other NW growth methods, such as chemical vapour deposition. In this context, the hydrothermal growth of ZnO NWs on seeded substrates immersed in equimolar zinc nitrate/HMTA aqueous solution was investigated. NWs were grown on polished silicon (001) substrates, and the solution concentrations, temperatures and growth times were varied. Importantly, the NW diameter was found to depend only on concentration during hydrothermal growth for times up to 4 hours. The diameter had an average value of 14 nm in 0.005 M solution and this increased up to a maximum of 150 nm in 0.07 M, when the NWs formed a continuous polycrystalline film. Concentration and temperature were both found to affect the axial growth rate of NWs in the [0001] direction. The growth rate was constant up to 4 hours (200 nm hr^{-1}) for constant conditions (81 °C, 0.025 M). The growth rate as a function of concentration was 7840 nm hr^{-1} M^{-1} up to 0.06 M (81 °C solution). The growth rate was a linear function of temperature with a value of 4.9 nm hr^{-1} K^{-1} (0.025 M solution). This indicates that growth takes place close to the equilibrium point, found by linear regression to be 36 °C for 0.025 M solution.

INTRODUCTION

It has been reported that ZnO nanostructures have a higher piezoelectric constant than bulk material [1,2]. ZnO is also a wide bandgap (3.37 eV) II-VI semiconductor with a large exciton binding energy (60 meV). Therefore, it has been investigated for the fabrication of short-wavelength light emitting diodes [3], varistors and transparent semiconductors. ZnO NWs have also been shown to exhibit lasing [4] and have been investigated for application in solar cells [5].

A low temperature hydrothermal growth process is advantageous for commercial manufacture because of its low thermal budget, and hence low cost compared to other growth techniques. Understanding the hydrothermal growth process, so as to exert control over the morphology of grown NWs, is therefore essential in order to commercialize functioning devices that exploit fully the physical properties of ZnO NWs. In this context, the effects of modifying the hydrothermal growth parameters (substrate, solution concentration, growth time and temperature) were investigated.

EXPERIMENTAL DETAILS

The hydrothermal growth method utilized was similar to that employed for microcrystal growth by Vergés et al. [6] and applied to NW growth by Greene et al. [7]. A 0.01 M seed solution was created by dissolving zinc acetate dihydrate ($Zn(CH_3COO)_2(H_2O)_2$, M_w=219.5 g) in propan-1-ol. The standard sample substrate was a polished (001) silicon wafer. The seed solution was spin-coated three times (2000 rpm for 30 s) onto the substrate. Between spinning, the sample was annealed on a hot plate at 120 °C for 1 minute, then cooled in ambient. Subsequently, samples were annealed in a furnace at 350 °C for 20 minutes to produce ZnO seeds on the surface (average diameter ~20 nm) through the reaction

$$2Zn(CH_3COO)_2 + 7O_2 \rightarrow 2ZnO + 8CO_2 + 6H_2O \qquad (1)$$

An equimolar solution of zinc nitrate hydrate ($Zn(NO_3)_2 \cdot 6(H_2O)$, M_w=297 g) and hexamethylenetetramine (HMTA, $C_6H_{12}N_4$, M_w=140.2 g) in DI water was prepared. The substrates were attached to Teflon supports placed in small flasks containing 20 ml of solution in an oven. It took 20 minutes for the solution temperature to stabilize. Although the flasks were left open, only 5% of the solution was found to evaporate at 92 °C during the 2 hour growth time. The ZnO seed layer acts as a focus for crystallization, ensuring dense growth of ZnO NWs. After growth, the samples were removed from the solution, rinsed with DI water, and dried with nitrogen. Figure 1 shows transmission electron microscopy (TEM) images confirming that the resulting NWs are single crystals.

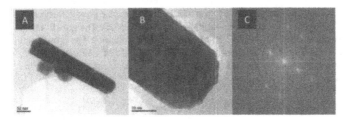

Figure 1. A) Top-view TEM of a single ZnO NW. **B)** Expanded view of part of ZnO NW. **C)** TEM Fourier Transform pattern.

Samples were cleaved and side view micrographs were made for each of five samples using an FEI Philips XL30 sFEG scanning electron microscope (SEM). The mean and standard deviation of the NW diameters and axial growth rates in the [0001] direction were determined from these micrographs.

The effect of varying the concentration of the equimolar zinc nitrate/HMTA growth solution, growth temperature and growth time was investigated, and each will be considered in the next section. During investigation of concentration and growth time, the solution temperature was maintained at 81 °C. During investigation of temperature and growth time, the concentration of growth solution was 0.025 M. The default growth time was 2 hours.

DISCUSSION

Concentration

As shown in Figure 2, the NW diameter changes with increasing concentration in a non-linear fashion, the increase being close to exponential at first, until a maximum value of ~ 160 nm is reached at around 0.07 M. Comparing Figures 2C and 2D leads to the conclusion that steric hindrance prevents further horizontal expansion of NWs.

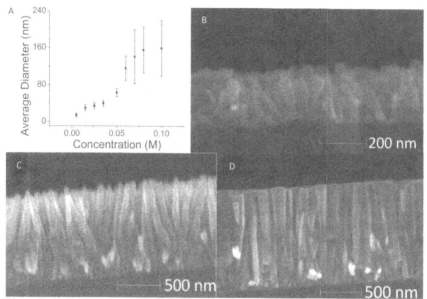

Figure 2. A) Graph of average diameter against concentration. The standard deviation is shown by the error bars. **B)** SEM micrograph of NWs grown in 0.015 M solution. **C)** SEM micrograph of NWs grown in 0.05 M solution. **D)** SEM micrograph of NWs grown in 0.07 M solution.

The axial growth rate of NWs is at first approximately linear with concentration, increasing at a rate of 7840 nm M^{-1} hr^{-1}, as shown in Figure 3. This is in agreement with previous research [8]. However, at 0.06 M the growth rate reaches a maximum of 470 nm hr^{-1}. This unexpected result demonstrates that the rate of inclusion of ions into the growing crystalline ZnO NW is not dependent only on the concentration of ions at the growth surface.

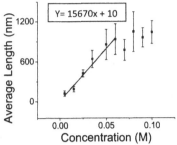

Figure 3. Graph of average length against concentration. The standard deviation is shown by the error bars.

It is unlikely that any of the observed changes were due to pH variation, since, when measured in liquids at room temperature, this was found to only reduce from 6 down to 5.8 as concentration increased. This is likely due to the buffering effect of the HMTA.

Time

The axial growth rate observed was completely linear at 200 nm hr^{-1} over a range of up to 4 hours (Figure 4), and yielded nanowires with an average diameter of 34 nm. Some previous work has suggested different growth rates between times less than one hour and greater than one hour [9]. However, this was not supported by our data, which showed a completely linear increase in nanowire length from 30 minutes until 4 hours. Further work has shown that the growth rate does begin to decrease after ~ 8 hours. This result is important as it demonstrates negligible change in reactant concentrations over the time period in which growth was carried out.

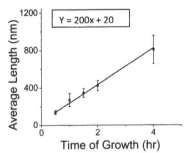

Figure 4. Graph of average length against time of growth. The standard deviation is shown by the error bars.

Temperature

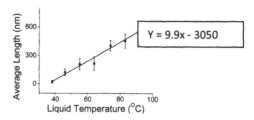

Figure 5. Graph of average length against solution temperature. The standard deviation is shown by the error bars.

At all temperatures, the NW diameter was found to remain constant, averaging 34 nm. The axial growth rate was observed to increase linearly with temperature. This is in agreement with theory, if we assume a thermally activated growth process with Arrhenius-type temperature dependence where the growth rate is given by

$$G = C_{in} e^{-\frac{Ein}{kT}} - C_{out} e^{-\frac{Eout}{kT}}$$

$$(2)$$

where E_{in} and E_{out} are the activation energies for inclusion of molecules from the liquid into the growing crystalline NWs, and the backward reaction into the liquid respectively. C_{in} and C_{out} are the concentration-dependent rate constants for these forward and reverse reactions. However, the preceding section demonstrated that these are constant over the two hour growth period. Therefore, linear growth is expected close to equilibrium (at $T=T_0$ no NW growth is expected) and so,

$$G \cong \gamma(T - T_0)$$

$$(3)$$

From Figure 5, γ can be determined to be 4.9 nm hr^{-1} K^{-1}. The equilibrium temperature (T_0) is found by linear regression to be 36 °C.

CONCLUSION

Many of the factors determining growth of ZnO NWs in zinc nitrate/HTMA aqueous solution were investigated to gain a more accurate understanding of, and control over, the growth process. Importantly, concentration was the only parameter found to have an effect upon the NW diameter, which increased with increasing concentration until steric effects prevented further expansion. The growth rate increased with concentration, up to a madximum growth rate when the NWs formed a continuous film. Interestingly, growth rate was found to be constant over time up to 4 hours and to increase linearly with temperature. This linear variation of growth rate with temperature demonstrates that hydrothermal growth of NWs takes place close to the equilibrium temperature, which is found by linear regression to be 36 °C for 0.025 M solution.

ACKNOWLEDGEMENTS

The authors would like to thank Mrs. Kalyan Sarma and Takeshi Kasama for help in collection of TEM data and Dr. Paul Beecher for help with editing. Research was supported by Nokia Research Centre Cambridge UK through the NRC/Cambridge University Framework Research Agreement Project on Large Area Sensing Surfaces. Michael E. Swanwick would like to thank the Gates-Cambridge Trust for their PhD funding.

REFERENCES

[1] M. Zhao, Z. Wang, and S.X. Mao, "Piezoelectric Characterization of Individual Zinc Oxide Nanobelt Probed by Piezoresponse Force Microscope," *Nano Letters*, vol. 4, Apr. 2004, pp. 587-590.

[2] J. Song, J. Zhou, and Z.L. Wang, "Piezoelectric and Semiconducting Coupled Power Generating Process of a Single ZnO Belt/Wire. A Technology for Harvesting Electricity from the Environment," *Nano Letters*, vol. 6, 2006, pp. 1656-1662.

[3] A. Nadarajah, R.C. Word, J. Meiss, and R. Konenkamp, "Flexible Inorganic Nanowire Light-Emitting Diode," *Nano Letters*, vol. 8, Feb. 2008, pp. 534-537.

[4] J.C. Johnson, H. Yan, R.D. Schaller, L.H. Haber, R.J. Saykally, and P. Yang, "Single Nanowire Lasers," *The Journal of Physical Chemistry B*, vol. 105, Nov. 2001, pp. 11387-11390.

[5] M. Law, L.E. Greene, J.C. Johnson, R. Saykally, and P. Yang, "Nanowire dye-sensitized solar cells," *Nat Mater*, vol. 4, Jun. 2005, pp. 455-459.

[6] M.A. Verges, A. Mifsud, and C.J. Serna, "Formation of rod-like zinc oxide microcrystals in homogeneous solutions," *Journal of the Chemical Society, Faraday Transactions*, vol. 86, 1990, pp. 959-963.

[7] L.E. Greene, B.D. Yuhas, M. Law, D. Zitoun, and P. Yang, "Solution-Grown Zinc Oxide Nanowires," *Inorganic Chemistry*, vol. 45, 2006, pp. 7535-7543.

[8] Y. Nakamura, "Solution-Growth of Zinc Oxide Nanowires for Dye-Sensitized Solar Cells," *NNIN REU 2006 Research Accomplishments*, NNIN, 2006.

[9] M. Guo, P. Diao, and S. Cai, "Hydrothermal growth of well-aligned ZnO nanorod arrays: Dependence of morphology and alignment ordering upon preparing conditions," *Journal of Solid State Chemistry*, vol. 178, Jun. 2005, pp. 1864-1873.

Mater. Res. Soc. Symp. Proc. Vol. 1174 © 2009 Materials Research Society   1174-V07-03

Dopants in Nanoscale ZnO

Matthew D. McCluskey,[1] Win Maw H. Oo,[1] and Samuel Teklemichael[1]

[1]Washington State University, Pullman, WA 99164-2814

ABSTRACT

Zinc oxide (ZnO) is a metal-oxide semiconductor that has attracted resurgent interest as an electronic material for a range of device applications. In our work, we have focused on how defect properties change as one goes from the bulk to the nanoscale. Infrared (IR) reflectance spectra of as-grown and hydrogen-annealed ZnO nanoparticles were measured at near-normal incidence. The as-grown particles were electrically semi-insulating, and show reflectance spectra characteristic of insulating ionic crystals. Samples annealed in hydrogen showed a significant increase in electrical conductivity and free-carrier absorption. A difference was observed in the *reststrahlen* line shape of the conductive sample compared to that of the as-grown sample. In addition to hydrogen doping, we successfully doped ZnO nanoparticles with Cu. To probe the electronic transitions of Cu^{2+} impurities in ZnO nanoparticles, IR transmission spectra were taken at liquid-helium temperatures. Two absorption peaks were observed at energies of 5781 and 5821 cm^{-1}. Finally, we tentatively assign a series of IR spectral lines to Na acceptors.

INTRODUCTION

Doping issues in nanocrystals are only beginning to be understood [1]. These issues are especially important in ZnO, a wide-bandgap metal-oxide semiconductor [2] that has attracted tremendous interest as a blue light emitting material [3], a buffer layer for GaN-based devices [4] and a transparent conductor [5] in solar cells [6]. Theoretical work has predicted ferromagnetism above room temperature for Mn-doped ZnO (given a large hole concentration) [7], an important requirement for spintronic devices. In addition to the potentially high Curie temperature (T_c), ZnO has numerous properties that are desirable for device applications, including low cost, environmental friendliness, and efficient light output.

In bulk, single-crystal ZnO doped with hydrogen or deuterium, infrared (IR) spectroscopy and Hall-effect measurements show that hydrogen binds to a host oxygen atom and donates an electron to the conduction band. The IR-active complex is unstable, decaying over a few weeks at room temperature. A more stable hydrogen donor may be substitutional hydrogen, predicted by Janotti and Van de Walle [8].

The effect of hydrogen on ZnO nanoparticles is more dramatic than for bulk crystals. Due to the large surface-to-volume ratio, hydrogen increases the free-carrier concentration at relatively low diffusion temperatures (300°C).

EXPERIMENTAL DETAILS

Nanoparticles were produced by the chemical reaction of zinc acetate dihydrate with sodium hydrogen carbonate, in an open-air furnace, at 200 °C for 3 hr. In prior work, x-ray diffraction (XRD) and transmission electron microscopy (TEM) results showed that the particles are 15-20 nm diameter with the hexagonal wurtzite structure [14]. The particles were pressed into 7 mm diameter pellets with a thickness of 0.25 mm. Some of the pellets were sealed in a quartz ampoule filled with 2/3 atm hydrogen and annealed at 350 °C for 1 hr [9].

IR reflectance spectra were obtained using a near-normal reflectance geometry, in a Bomem DA8 Fourier transform infrared (FTIR) spectrometer. IR spectra between 200 and 500 cm^{-1} were obtained with a Mylar beam splitter and DTGS detector. A KBr beamsplitter and MgCdTe detector were used for wavenumbers above 500 cm^{-1}. A gold mirror was used as a reference. Transmission spectra were obtained for ZnO nanopowder mixed in a KBr pellet.

DISCUSSION

Hydrogen-doped ZnO nanoparticles

The effect of hydrogen on the conductivity of ZnO nanoparticles has implications for nanoscale optoelectronic devices. In this study, infrared (IR) reflectance spectra of as-grown and hydrogen-annealed ZnO nanoparticles were measured at near-normal incidence. The as-grown particles were electrically semi-insulating, and show reflectance spectra characteristic of insulating ionic crystals. Samples annealed in hydrogen showed a significant increase in electrical conductivity and free-carrier absorption. A difference was observed in the *reststrahlen* line shape of the conductive sample compared to that of the as-grown sample. The effective medium approximation was applied to model the reflectance and absorption spectra. The agreement between experimental results and the model suggests that the nanoparticles have inhomogeneous carrier concentrations.

IR reflectance spectra are shown in Fig. 1 for an as-grown sample and a sample annealed in hydrogen. The reflectivity of the as-grown sample shows a sharp decrease in reflectivity ($R_{min} \approx$ 0) near the longitudinal optical phonon frequency ω_{LO}. This feature is a typical *reststrahlen* band reflection of semi-insulating ZnO [7]. The hydrogen annealed sample exhibits a change in the *reststrahlen* line shape, and the disappearance of R_{min} was observed. The reflectance change is consistent with an increase in the free carrier concentration. To model the reflectance spectra, the dielectric function of ZnO was calculated by a classical Lorentz-Drude model. To obtain fits to the data (dotted lines), we had to assume an inhomogeneous distribution of doping among the nanoparticles. Some nanoparticles were heavily doped ($n \sim 10^{19}$ cm^{-3}) while others were lightly doped ($n \sim 10^{17}$ cm^{-3}).

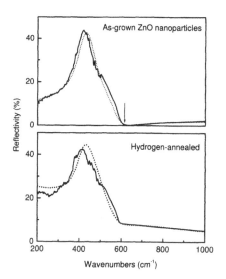

Figure 1. IR reflectance spectra of ZnO nanoparticles. The dotted lines denote the calculated spectra.

ZnO:Cu nanoparticles

To probe the electronic transitions of Cu^{2+} impurities in ZnO nanoparticles, IR transmission spectra were taken at liquid-helium temperatures. Two absorption peaks were observed at energies of 5781 and 5821 cm^{-1} (Fig. 2). These absorption peaks arise from internal transitions of Cu^{2+} ions from the two sublevels of the E state to the lowest T_2 level. Similar absorption lines were observed in bulk ZnO [10,11]. However, the width of the absorption peaks in our nanoparticles is broader than that of bulk crystals, perhaps due to inhomogeneous strains and/or electric fields.

Unidentified IR spectrum: Na acceptors?

ZnO nanoparticles grown using sodium hydrogen carbonate showed a series of IR peaks at low temperatures (Fig. 3). These peaks are consistent with an acceptor with a hole binding energy of ~ 0.4 eV. Recent experimental work has shown that alkali atoms diffused into bulk ZnO result in 'shallow' acceptors with a hole binding energy of 0.3 eV [12]. Hence, we tentatively conclude that some of the nanoparticles contain a Na acceptor atom, which substitutes for a Zn atom. If this turns out to be the case, then Na doping could be feasible for synthesizing p-type ZnO nanoparticles.

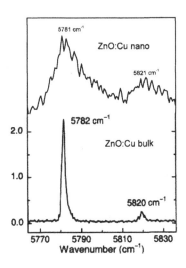

Figure 2. IR spectra of ZnO:Cu nanoparticles (this work) and bulk crystal (Ref. 10, courtesy of L. Halliburton).

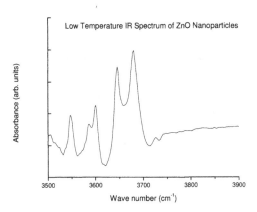

Figure 3. IR spectrum of ZnO nanoparticles, showing peaks that may be due to Na acceptors.

CONCLUSIONS

We have studied ZnO nanoparticles doped with H and Cu, as well as an unknown dopant that we tentatively attribute to Na. Future studies will determine the effect of nanoparticle size on doping efficiency and properties.

ACKNOWLEDGMENTS

This work was supported by the U.S. Department of Energy under Grant No. DE-FG02-07ER46386 and the U.S. National Science Foundation under Grant No. DMR-0704163.

REFERENCES

1. D.J. Norris, A.L. Efros, and S.C. Erwin, Science **319**, 1776 (2008).
2. Ü. Özgür, Ya. I. Alivov, C. Liu, A. Teke, M.A. Reshchikov, S. Dogan, V. Avrutin, S.-J. Cho, and H. Morkoç, J. Appl. Phys. **98**, 041301 (2005).
3. D.C. Look, Mater. Sci. Engin. B **80**, 383 (2001).
4. R.J. Molnar, in Semiconductors and Semimetals **57** (Academic Press, New York, 1999), pp. 1-31.
5. T. Minami, MRS Bulletin **25** (8), 38 (2000).
6. A. Nuruddin and J.R. Abelson, Thin Solid Films **394**, 49 (2001).
7. T. Dietl, H. Ohno, F. Matsukura, J. Cibert, and D. Ferrand, Science **287**, 1019 (2000).
8. A. Janotti and C.G. Van de Walle, Nature Materials **6**, 44 (2007).
9. W.M. Hlaing Oo, M.D. McCluskey, J. Huso, and L. Bergman, J. Appl. Phys. **102**, 043529 (2007).
10. R.E. Dietz, H. Kamimura, M.D. Sturge, and A. Yariv, Phys. Rev. **132**, 1559 (1963).
11. N.Y. Garces, L. Wang, L. Bai, N.C. Giles, L.E. Halliburton, and G. Cantwell, Appl. Phys. Lett. **81**, 622 (2002).
12. B.K. Meyer, J. Stehr, A. Hofstaetter, N. Volbers, A. Zeuner, and J. Sann, Appl. Phys. A **88**, 119 (2007).

Mater. Res. Soc. Symp. Proc. Vol. 1174 © 2009 Materials Research Society   1174-V05-02

The Challenge for Large-Scale Vapor-Phase Growths of Not-Catalyzed ZnO Nanostructures: Purity vs. Yield

Davide Calestani, Ming Zheng Zha, Roberto Mosca, Laura Lazzarini, Giancarlo Salviati, Andrea Zappettini, Lucio Zanotti
IMEM-CNR, Parco Area delle Scienze 37/A, Parma, PR-43124, ITALY

ABSTRACT

Large-scale growth capability is a general requirement for any reliable and cost-effective device application. Catalyst-free vapor-phase growth techniques generally let obtain high purity materials, but their application in large-scale growths of zinc oxide (ZnO) nanostructures is not trivial, because the lack of catalysts makes the control of these process rather difficult. Three different optimizations of the basic vapor phase growth have been studied and performed to obtain selected and reproducible growths of three different ZnO nanostructures with improved yield, i.e. nanotetrapods, nanowires and nanorods. No precursor or catalyst has been used in order to reduce contamination sources as more as possible.

INTRODUCTION

ZnO nanostructures are today a very important research topic because their proved (or even just "potential") properties promoted huge studies in many different application fields, such as optoelectronics, photovoltaics, spintronics, gas sensing, photocatalysis, piezo-electric applications, etc. (see Ref. [1-6] for a general overview). Since a reproducible large-scale production is essential for a likely use of these nanostructures in any industrial device or application, large efforts have been done to control and stabilize their synthesis processes.

Good results have been obtained in vapor phase growths of nanorods and nanowires, by mean of metal catalysts (such as Au, Pt or Ni particles) [7-10]. On the other side, large and controlled production of some ZnO nanostructures have been realized by wet chemical processes (e.g. see Ref. [11-17]). Unfortunately both these approaches are intrinsically affected by the introduction of impurities in the nanocrystals' structure. Even if sometimes the presence of such impurities is negligible, they often have a strong effect on the physical properties of these semiconducting nanostructures.

Catalyst-free vapor-phase growth techniques should not be affected by the same high impurity levels if high purity sources and gases are employed. Unfortunately, the synthesis control is generally more difficult in this kind of processes.

In the present work authors show the results obtained in the optimization of three different growth processes, for a large-scale oriented production of (i) ZnO tetrapods, (ii) ZnO long nanowires and (iii) ZnO nanorods. All the described processes share a catalyst-free growth and the use of high purity metallic Zn, O_2 and inert carrier gas (Ar) only. The respective growth mechanisms and the possible improvements in process yield are discussed.

Optimized vapor phase synthesis processes for ZnO tetrapods, nanowires and nanorods were performed by modifying some growth detail inside the same reactor, which is a tubular furnace with a quartz tube through which different gases can be flowed (figure 1). Source material was common for all these processes and it is made of metallic Zn foils (purity 99.999%), which are rapidly etched in diluted acid before the use to remove oxide traces that might reduce the evaporation rate during the growth process. Then different parameters were chosen to obtain the three mentioned nanostructures.

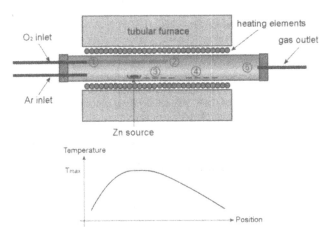

Figure 1. Scheme of the growth reactor were synthesis processes were performed. For ZnO nanowires the tube for O_2 inlet is in position (1) and substrates in position (3); for ZnO nanorods the tube for O_2 inlet is in position (1) and substrates in position (4); for ZnO tetrapods the tube for O_2 inlet is in position (2) and collection zone is in position (5). At bottom a sketch of the growth profile is reported.

ZnO tetrapods (figure 2a) were grown directly in the vapor phase. Source material was placed in the hottest zone of the furnace and then heated up to 700°C in an inert gas flow (Ar, 100 sccm). Zn vapor was carried by the inert Ar flow along the furnace tube up to when it matched the oxygen flow (10 sccm), which enter the reaction tube in a second zone (marked with label "2" in figure 1), where temperature was set in the range 600-500°C, to start tetrapods nucleation and growth. While floating in the gas stream, tetrapods grew until they reached the cold zone at the end of the furnace (marked with label "5" in figure 1). In this zone they arrived in form of a visible and continuous white smoke and they deposited on walls of the quartz tube, forming a thick and fluffy layer. The growth of a single tetrapod takes place in less than 2 minutes but, since Zn source was protected from oxidation by the inert gas flow, the whole synthesis process could run up to complete source exhaustion.

ZnO nanowires (figure 2b), instead, were obtained directly on alumina substrates, where a 10 μm Zn layer was previously deposited. Source material was heated up to 650°C in a 20 sccm Ar flow and substrates were positioned downstream, close to the source (position "3" in

46

figure 1) where temperature is 20-30°C lower. When this temperature was attained, O_2 flow is added (Ar/O_2 ratio was 5:1) to allow ZnO nanowire formation. These conditions were maintained for 30 minutes and then the furnace was cooled down to room temperature.

ZnO nanorods (figure 2c) were obtained similarly to nanowires on substrates with a pre-deposited Zn layer, but in this case thickness was 2 μm. Source material was heated at 600°C in a 50 sccm Ar flow and substrates were placed downstream in a colder region (position "4" in figure 1), at about 480-450°C, just before the position in which Zn vapor strongly condensed (around melting point).

Figure 2. SEM images of grown ZnO nanostructures: a) nanotetrapods, b) long nanowires, c) nanorods.

All the obtained samples were characterized by Scanning Electron Microscope (Philips SEM 515), to investigate their morphology, and by X-Ray Diffraction (Thermo ARL X'tra XRD), to study their crystalline structure and to exclude the presence of residual metallic Zn.

DISCUSSION

ZnO tetrapods, nanowires and nanorods have been obtained with different processes inside the same reactor. High material purity is obtained by using only high purity Zn, O_2 and Ar. Sample characterization by SEM and XRD showed that the obtained nanostructures are rather homogeneous in dimension, well crystallized and that no traces of residual metallic Zn is present among them (figure 3).

All these three ZnO nanostructures grow in very high supersaturation. However, different conditions are required for the growth of specific nanostructures. Specific optimizations were done in the growth set-up in order to maintain these particular conditions during the whole process. In this way, synthesis yield was strongly improved without affecting material purity.

ZnO tetrapods

ZnO tetrapods generally nucleates and growth while floating in the gas stream, so high supersaturation has to be reached in the vapor phase, keeping Zn vapor and O_2 pressures high and comparable (in the order of several millibars, in our growth system). In standard growth processes described in literature oxygen is introduced directly in the growth reactor where Zn source evaporates. This approach supplies the correct Zn and O_2 pressures only for a limited time, because oxygen introduction also starts source oxidation and, consequently, a continuous decrease of Zn evaporation, up to complete reaction quenching. These processes are strongly self-limited.

The main aim of the optimized synthesis process we described is to prevent source oxidation and to keep Zn evaporation flux constant during the whole process, so that ZnO tetrapods can continuously nucleate and grow up to complete source exhaustion, i.e. with a much higher yield.

This goal has been attained by leaving Zn to effuse from its container under an inert Ar flow, which carries Zn vapors to a further reactor zone where oxygen is introduced and ZnO tetrapod growth starts. Growing tetrapod is then carried by the flow up to the end of the furnace, where it deposits on the walls of the reactor tube. This streaming growth allows key-processes of the growth mechanism to take place in stationary conditions, at fixed positions in the reactor, so that ZnO tetrapod synthesis takes place like in an assembly line. In our small laboratory-scale reactor it has been possible to grow, in this way, up to 1g of tetrapods in a single run, which corresponds to about 10^{11}-10^{13} tetrapods.

Legs of obtained tetrapods have an average thickness is in the range 20-200nm, while their length is generally a few micrometers. They can be easily removed from the end of the furnace in the form of a thick and fluffy film. This material can be easily deposited by alcoholic suspension but, if required by final application, it may be also used for the preparation of screen-printing inks.

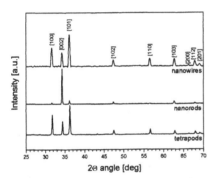

Figure 3. X-ray diffraction patterns for ZnO tetrapods, nanowires and nanorods. All the peaks can be indexed in the common hexagonal wurtzite cell (P63mc, with a=3.25 and c=5.20). Tetrapods and nanowires have been measure with grazing-incidence geometry, to avoid substrate contribution. The pattern of nanorods clearly shows a preferential orientation for [001]. No peak for metallic Zn is present.

ZnO nanowires and nanorods

On the other side, it has been observed that ZnO nanowire growth is favored when liquid Zn is present on the substrate. Unfortunately it is difficult to obtain Zn condensation in the high temperature zone of the furnace (600-650°C), where Zn equilibrium vapor pressure is high (about 10^{-2} bar) and evaporation process is strongly competitive.

If no catalyst is used, Zn condenses only occasionally in the roughest zones of the substrates, where Zn micro-droplets nucleation is favored. This random and not reproducible process can be stabilized if a Zn layer is deposited on the substrates before the growth. This layer plays a double role, i.e. it directly furnishes liquid Zn to promote the growth of nanowires and, at the same time, it creates a stable region close to the substrate surface where Zn vapor pressure is kept constant by continuous feeding from the liquid phase.

In this case, layer thickness becomes very important because it must be high enough to grant the feeding of nanowires during the growth process and to compensate Zn vapor losses due to the natural evaporation, but at the same time not in excess to avoid the presence of metallic Zn residuals below the nanowires at the end of the growth. In the described process and in our laboratory reactor, a 10 µm thick layer gives the best results. Even if a rather thick layer is used to compensate the large Zn evaporation, an additive Zn source is still positioned in the zone of the furnace where temperature is maximum (650°C) in order to fill reactor with Zn vapor and to reduce even more early evaporation of Zn from the layers on the substrates.

The obtained system grants a "buffered" Zn rich zone close to the surface of substrates and allowed us to extend the growth region, in our reactor, up to several square centimeters, where ZnO nanowires grow homogeneously and reproducibly. Their average diameter is 10-50nm, while they generally develop in length up to tens or hundreds of microns.

Also ZnO nanorods have been obtained by the use of a pre-deposition of Zn metal layer on the substrates, but different conditions are requested to grow nanorods instead of nanowires. Nanorods have been homogeneously obtained on a few square centimeters area by depositing a 2 µm thick layer on the substrates and placing them downstream at about 450°C.

In the substrate region Zn vapor pressure is rather low (about 10^{-4} bar) and temperature is just above Zn melting point. In these conditions high supersaturation is obtained only in a very narrow region above the liquid layer and the growth of these linear nanocrystals is limited to this short range.

The obtained ZnO nanorods are very homogeneous in size and, depending on the growth conditions, their thickness ranges between 30 nm and 100 nm. In this kind of growth final length of the nanorods seems to be correlated with film thickness. Unfortunately the growth of nanorods on layers thicker than 2 µm was unsuccessful and, in that case, some metal remained below the nanorods at the end of the growth.

CONCLUSIONS

Catalyst-free vapor phase growths are generally preferable when high purity is strongly required, but unfortunately yield and reproducibility of these techniques are not as good as those obtained with wet processes. Nevertheless, selectivity, reproducibility and yield in the preparation of ZnO tetrapods, nanowires and nanorods can be strongly improved by specific optimizations of the respective growth processes, which have been here described, without affecting material purity. A large quantity of tetrapods can be obtained by a streaming growth along the reactor, where constant flow of Ar, Zn vapor and O_2 are maintained during the whole process. On the other side selective growth of nanowires and nanorods can be obtained, with different growth conditions, on several square centimeters of different substrates if a Zn film of the proper thickness is pre-deposited on them.

All these nanostructures have been obtained in the same growth reactor. Since good results have been achieved in a small laboratory-scale reactor, even better results are expected by the scale-up of the growth system.

ACKNOWLEDGMENTS

This work has been partially supported by the Regional Laboratories "PROMINER" and "ENVIREN" and by CARITRO Foundation in the frame of "DAFNE" Project.

REFERENCES

1. J. G. Lu, P. Chang, Z. Fan, *Materials Science and Engineering R* **52**, 49-91 (2006).
2. L. Schmidt-Mende, J. L. MacManus-Driscoll, *Materials Today* **10**, 40-48 (2007).
3. Z. L. Wang, *J. Phys.: Condens. Matter* **16**, R829-R858 (2004).
4. C. Klingshirn, *Phys. Stat. Sol. B* **244**, 3027-3073 (2007).
5. Ü. Özgür, Ya. I. Alivov, C. Liu, A. Teke, M. A. Reshchikov, S. Doğan, V. Avrutin, S.-J. Cho, H. Morkoç, *J. Appl. Phys.* **98**, 041301 (2005).
6. M. A. Arnold, Ph. Avouris, Z. W. Pan, Z. L. Wang, *J. Phys. Chem. B* **107**, 659-663 (2003).
7. D. Banerjee, J Rybczynski, J. Y. Huang, D. Z. Wang, K. Kempa, Z. F. Ren, *Appl. Phys. A-Mater.* **80**, 749-752 (2005).
8. Z. M. Zhu, T. L. Chen, Y. Gu, J. Warren, R. M. Osgood, *Chemistry of Materials* **17**, 4227-4234 (2005).
9. S. W. Kim, S. Fujita, *Applied Physics Letters* **86**, 153119 (2005).
10. P. X. Gao, Y. Ding, I. L. Wang, *Nano Letters* **3**, 1315-1320 (2003).
11. R. Viswanatha, P. K. Santra, C. Dasgupta, D. D. Sarma, *Physical Review Letters* **98**, 255501 (2007).
12. U. Pal, P. Santiago, *J. Phys. Chem. B* **109**, 15317-15321 (2005).
13. K. Elen, H. Van den Rul, A. Hardy, M. K. Van Bael, J. D'Haen, R. Peeters, D. Franco, J. Mullens, *Nanotechnology* **20**, 055608 (2009).
14. T. Mahalingam, K. M. Lee, K. H. Park, S. Lee, Y. Ahn, J. Y. Park, K. H. Koh, *Nanotechnology* **18**, 035606 (2007).
15. H. X. Zhang, J. Feng, J. Wang, M. L. Zhang, *Materials Letters* **61**, 5202-5205 (2007).
16. J. M. Jang, S. D. Kim, H. M. Choi, J. Y. Kim, W. G. Jung, *Mater. Chem. Phys.* **113**, 389-394 (2009).
17. L. E. Greene, B. D. Yuhas, M. Law, D. Zitoun, P. D. Yang, *Inorganic Chemistry* **45**, 7535-7543 (2006).

Mater. Res. Soc. Symp. Proc. Vol. 1174 © 2009 Materials Research Society   1174-V07-01

A Facile Method for Patterning Substrates With Zinc Oxide Nanowires

Jeong-Hyun Cho[1], Elizabeth W. Cha[2], and David Gracias[1,3]

[1] Department of Chemical and Biomolecular Engineering, Johns Hopkins University, Baltimore, MD 21218, U.S.A.

[2] Department of Biomedical Engineering, Johns Hopkins University, Baltimore, MD 21218, U.S.A.

[3] Department of Chemistry, Johns Hopkins University, Baltimore, MD 21218, U.S.A.

ABSTRACT

Conventional growth of zinc oxide (ZnO) nanowires (NWs) is typically carried out using vapor liquid solid (VLS) and chemical vapor deposition (CVD) methods. While these methods are effective, they often involve the use of specialty gasses and equipment. We have discovered that ZnO NWs grow spontaneously from zinc (Zn) films (thermally evaporated on silicon (Si) substrates) when the films are merely heated on a hot-plate in air at ambient pressures for 10 minutes. This process does not involve any metal catalysts, seed layers, specialty gasses or surface treatments in forming patterned regions (on silicon substrates) of NWs with typical diameters in the range of 20-50 nm and lengths of 2-3 μm.

INTRODUCTION

As a material, bulk zinc oxide (ZnO) is a wide energy band gap semiconductor with a band gap of 3.37 eV and a large exciton binding energy of 60 meV [1]. Additionally, ZnO crystals show piezoelectric properties. ZnO is used in a large variety of products ranging from sunscreen to light emitting diodes (LED). On the nanoscale, ZnO structures possess electrical, chemical, mechanical and optical properties giving them the capability to improve current devices and aid in the development of new technologies.

Conventional growth of ZnO nanowires (NWs) is carried out using vapor liquid solid (VLS) and chemical vapor deposition (CVD) methods. These methods typically require specialty gasses, a catalyst and sometimes specialty reaction vessels. Catalytic or seed layer impurities can alter the electronic properties of ZnO NWs and hence these catalytic methods are not ideal, especially when the NWs are used to construct electronic or optical components such as diodes or resonators.

There are some methods to grow ZnO NWs on silicon (Si) substrates without the use of any catalysts or seed layers but they involve the use of surface modification of the substrate such as by mechanical scratching, polishing, or chemical etching. It has been argued that the nucleation of ZnO on Si substrates is hard to achieve due to a large lattice mismatch between

ZnO and Si [2]. Hence, in order to grow NWs without the use of pre-nucleant materials, surface treatments are employed [3]. There are still other methods that do not utilize any catalysts, namely seed layers or surface modification, but these methods have either been shown to work with only bulk zinc pieces [4] or require high temperatures (1300 °C) [5]; these processes make it difficult to integrate ZnO NWs with conventional electronic fabrication processes.

Recently, we have discovered that ZnO nanowires grow spontaneously, when thermally evaporated Zn patterns deposited on commercial Si wafer substrates are heated in air. NW growth occurred without the need for any surface treatments, seed layers, catalysts, or specialty gasses. We believe that this method is simple, straightforward and cost-effective. The method also does not require any specialty equipment and allows facile integration of ZnO NWs in electronic devices.

EXPERIMENT

We utilized commercial 3 inch <100> silicon wafer substrates. Zn thin films were patterned using conventional photolithography and lift-off metallization (Fig. 1). Briefly, this process involved spin coating approximately 2μm of photoresist and thermal evaporation of 800 nm of Zn with 25 A°/sec at 1 x10⁻⁵ Torr. After deposition, the wafer was rinsed with acetone for approximately 5 minutes to strip the photoresist. ZnO NWs formed spontaneously when the patterned Zn substrates were heated on a hot plate. This heating was achieved by placing the substrates in the center of a hotplate. The temperature was ramped from room temperature to the desired temperature in approximately 2-5 minutes. The desired temperature was maintained for 10 minutes. After heating, the samples were cooled for 15 minutes and immediately imaged using a scanning electron microscope (SEM).

RESULTS and DISCUSSION

Figure 1 shows the schematic diagram of the process. The first step involved the deposition of pure Zn films onto Si wafer substrates. We observed that no Zn adhered to the Si

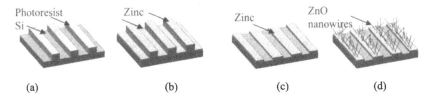

(a) (b) (c) (d)

Figure 1. Schematic diagram of the process used to pattern ZnO NWs (a) A commercial Si wafer was photolithographically patterned (b) Zn was thermally evaporated at a rate of approximately 25 A°/s (c) After dissolving the photoresist a Zn pattern is left behind (d) On heating to 500 °C for 10 minutes ZnO NW growth initiates.

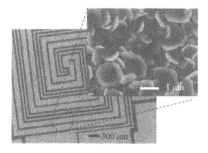
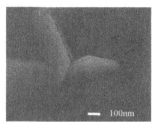

Figure 2. SEM images at different magnifications showing 800 nm thick Zn thin film patterns. The Zn films have a disc shaped morphology.

substrate when the deposition speeds ranged from 0.1 to 10 A°/sec. With a deposition speed of 15 A°/sec, Zn adhered only on top of the photoresist layer. With a deposition rate of 25 A°/sec, an 800 nm layer of Zn adhered not only on top of the photoresist but also on the Si surface. Hence, in order to allow for the deposition of Zn on the substrate, without an intermediate adhesion promoter or wetting layer, it was necessary to evaporate at speeds above 25 A°/s. We observed that the resulting Zn films formed had a disc shaped morphology (Fig. 2). The average diameter of the discs was approximately 1 μm with a thickness of approximately 10 nm.

When these patterned substrates were heated on a hot plate, NWs formed spontaneously. These wire shaped structures resemble whiskers; however Zn whiskers are typically several microns in diameter as opposed to sub-100 nm structures observed here.

We varied the heating temperature from 200° C - 600° C to study its influence on the NW growth (Fig. 3). In all cases, samples were heated for 10 minutes at which point the hotplate was switched off. After heating to 200° C, no NW growth was observed. The extent of NW growth

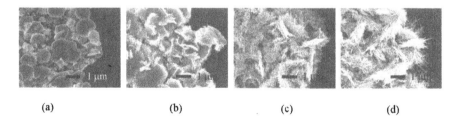

| (a) | (b) | (c) | (d) |

Figure 3. Top-view SEM images of patterned Zn films on silicon substrate after heating. The films were heated on a hotplate for 10 minutes at (a) 200° C (b) 300° C (c) 400° C and (d) 500° C

Figure 4 SEM images of ZnO nanowires grown on Zn film patterns.

increased with increasing temperature; additionally the NW length increased until a temperature of 500° C, but appeared to remain constant at higher than 600° C.

Additionally, when the heating time was increased from 10 to 60 minutes at 500 °C, no noticeable difference was observed. Hence, we recommend NW growth conditions at 500 °C for 10 minutes. Under these conditions, NWs form with typical diameters in the range of 20-50 nm and lengths of 2-3 μm (Fig. 4). In order to verify the composition of the NWs, we also performed energy dispersive spectroscopy (EDS). Through EDS of the Zn NW patterns we observed only peaks for Zn and oxygen (O) (Fig. 5).

We also observed interesting NW growth features including branching; this branching resulted in the formation of bridges (Fig. 6), which may be useful in interconnecting electronic devices.

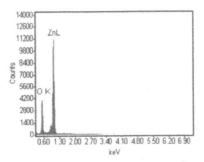

Figure 5. The EDS analysis of nanowires corresponding to Zn and O.

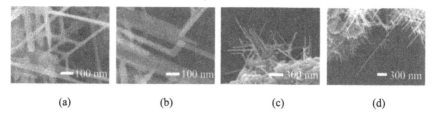

(a)   (b)   (c)   (d)

Figure 6. ZnO nanowires SEM image showing (a) a bridge, (b, c) a change in the direction, and (d) creation of NW branches in the middle of a nanowire.

CONCLUSIONS

In conclusion, we have demonstrated an easy way to form ZnO NWs by heating patterned Zn on Si substrates without the need for any catalysts, specialty gasses, seed layers, surface modification etc. The process was repeated several times and was reproducible under the conditions stated above. We are unclear of an exact mechanism for NW growth; the structures are reminiscent of Zn whiskers, but are sub-100 nm sized. Moreover, the process is very simple and allows easy lithographic patterning of NW areas as well as integration with electronic devices. It should be noted that although we are certain that the NW regions contain only Zn and O (as determined by EDS), more analysis is needed to determine the exact stoichiometric composition as well as the crystalline orientation of the NWs. It is possible, for example, that the NWs have a Zn metallic core with an oxide surface layer. Measurements of NW electronic properties as well as NW TEM analysis are future research directions that would elucidate these structural properties.

REFERENCES

1. K. Hümmer, *Phys. Status Solidi B* **56**, 249 (1973).
2. J. S. Jeong, J. Y. Lee, J. H. Cho, C. J. Lee, S. J. An, G. C. Yi, and R. Gronsky, *Nanotechnology* **16**, 2455 (2005).
3. S. T. Ho, C. Y. Wang, H. L. Liu, and H. N. Lin, *Chemical Physics Letters* **463**, 141 (2008).
4. S. Ren, Y.F. Bai, J. Chen, S.Z. Deng, N.S. Xu, Q.B. Wu, and Shihe Yang, *Materials Letters* **61**, 666 (2007).
5. Y.J. Xing, Z.H. Xi, X.D. Zhang, J.H. Song, R.M. Wang, J. Xu, Z.Q. Xue, and D.P. Yu, *Applied Physics A* **80**, 1527 (2005).

Mater. Res. Soc. Symp. Proc. Vol. 1174 © 2009 Materials Research Society   1174-V07-09

Characterization of ZnO Nanostructures Grown by Pulsed Laser Deposition

Christian Weigand[1], Matt Bergren[3], Cecile Ladam[2], Per Erik Vullum[2], John C. Walmsley[2], Ragnar Fagerberg[2], Tom Furtak[3], Reuben Collins[3], Jostein Grepstad[1] and Helge Weman[1]

[1]Department of Electronics and Telecommunications, Norwegian University of Science and Technology, NO-7491 Trondheim, Norway
[2]SINTEF Materials and Chemistry, NO-7465 Trondheim, Norway
[3]Department of Physics, Colorado School of Mines, Golden, CO 80401, U.S.A.

ABSTRACT

ZnO nanostructures were grown by pulsed laser deposition on c-plane sapphire substrates. The as-grown nanostructures were examined by scanning electron microscopy and transmission electron microscopy ZnO nanowires were grown using a gold catalyst, at a high substrate temperature of 800°C and an ambient gas pressure of 0.5 mbar (5% oxygen, 95% argon). Changing the gas composition to pure oxygen led to the growth of stacking fault-free ZnO nanosheets with their growth direction inclined to the [0001] direction. Similar nanosheets with stacking faults were found when lowering the growth temperature to 600°C for a 5% oxygen – 95% argon ambient gas composition and the same overall pressure. A growth mechanism for these ZnO nanosheets is proposed.

INTRODUCTION

ZnO has been subject to extensive research in recent decades. Due to its wide band gap of 3.37 eV and its large exciton binding energy of 60 meV at room temperature, it is especially interesting for applications in short wavelength light emitting devices and solar cells [1]. ZnO is abundant in nature, and thus a suitable material for low-cost devices. It can be grown by different deposition techniques, such as metal-organic chemical vapor deposition [2], wet chemical synthesis, and different chemical and physical vapor deposition techniques, including pulsed laser deposition (PLD) [3-5]. The latter technique allows for controlled growth of various materials, preserving the stoichiometry of the source material upon transport from target to substrate. In PLD, ambient gases such as oxygen and argon can be introduced in order to aid the growth of metal oxides. In this work, ZnO nanorods and nanosheets were grown on sapphire substrates at different temperatures and ambient gas compositions from a ZnO target. Scanning electron microscopy (SEM) and transmission electron microscopy (TEM) were used to investigate the growth mechanisms of these nanostructures for different growth conditions.

EXPERIMENT

The ZnO nanostructures were grown on c-plane sapphire substrates using a KrF excimer laser at 248 nm wavelength (Lambda Physik LPX Pro 210i). A raster-scanned ZnO target (American Elements, 99.999%) was ablated with a laser fluency of approximately 1 J/cm² at a pulse repetition rate of 10 Hz and a target-to-substrate distance of 45 mm. The base pressure of the vacuum chamber before deposition was less than 2×10^{-7} mbar. The substrate temperature was varied between 600 °C and 800 °C, and ambient background gases (oxygen and argon) were introduced into the growth chamber. The gas pressure during deposition was kept at 0.5 mbar, with a constant flow of either pure oxygen or a 5%:95% oxygen:argon ratio.

Prior to ZnO deposition, the substrates were cleaned in an ultrasonic bath with acetone and ethanol and then blown dry with pure nitrogen gas. Thereafter, a thin layer of Au with a nominal thickness of 2.5 nm was deposited in an electron beam evaporation chamber to create seeds for the growth. The Au-coated substrates were then glued onto a sample holder with thermally conductive silver paste and introduced into the PLD chamber.

After deposition, the samples were left to cool down to room temperature at a rate of about 15 °C/min, and then investigated by SEM (Zeiss Supra VP55) and TEM using a JEOL 2010F and a Philips CM30, both operating at 200 kV.

RESULTS AND DISCUSSION

Figure 1 shows SEM images of the ZnO nanostructures grown by PLD at two different substrate temperatures. The surface topography of a sample grown for 60 min at $T_s = 800°C$ in an ambient of 5% O_2 and 95% Ar is displayed in Figure 1a. Under these conditions, ZnO nanowires grow predominantly vertically from the substrate surface via the vapor-liquid-solid (VLS) mechanism, as indicated by the presence of Au particles at the nanowire tips [6]. The observed nanowires have a typical length of ~1 μm and an average diameter of about 60 nm with slight tapering towards the tip. Interestingly, when growing at a lower substrate temperature $T_s = 600$ °C with all other parameters constant, the growth mode changes from nanowires to planar and tapered, two-dimensional structures which we hereafter refer to as nanosheets (see Figure 1b). The nanosheets display an anisotropic thickness distribution, i.e. the width significantly exceeds the depth. We also note that despite the shorter growth time of 15 minutes, the length of the sheets was comparable to the length of the nanowires grown at $T_s = 800$ °C for one hour. Au particles present at the nanosheet tips suggest a VLS growth mechanism. However, the observed nanosheets exhibit rather large widths of about 100 nm at their base. Therefore, in order for these

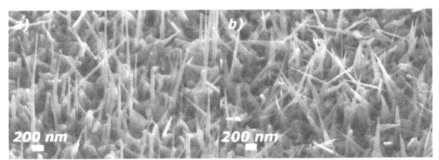

Figure 1: 45°-tilted view SEM images of ZnO nanostructures grown at 5% O_2 and 95% argon ambient gas. a) ZnO nanowires grown at $T_s = 800$ °C for 60 min, b) Predominantly ZnO nanosheets grown at $T_s = 600$ °C for 15 min.

structures to grow by the VLS mechanism only, the size of the Au droplet would have to be comparable in width to the sheet base. Thus, Au droplets of about 100 nm in diameter would be required. However, the thickness of the Au film deposited prior to ZnO growth was only about 2.5 nm. The size of the Au droplets after annealing at the growth temperature and before deposition of ZnO was determined by SEM (not shown) to be $10 - 20$ nm and thus, considerably smaller than the base widths of the ZnO sheets. As a consequence, we consider it unlikely that the Au droplets acting as seeds are the only factor responsible for the observed growth of nanosheets.

An interesting feature of these nanosheets is apparent from the TEM data. In particular, in the electron diffraction pattern shown in the inset of Figure 2e the growth direction of the nanosheets does not follow the preferred [0001] growth direction for ZnO nanostructures reported in literature [3, 7-10], but is inclined to this axis by about 27°. Moreover, numerous stacking faults are visible in the dark-field and the high-resolution TEM images of Figure 2d and e. These defects do not display a regular pattern, but seem to occur randomly along the [0001] direction. TEM images of these nanosheets (see Figure 2) also show that the sidewalls of the nanosheets have a rather rough surface morphology with numerous steps.

Even though the majority of the observed structures on this sample are identified as nanosheets, some nanowires are also found, as seen in the bright-field TEM image of Figure 2b. Contrary to the nanosheets, these wires exhibit no stacking faults, but are perfect single crystals with their growth axis parallel to the ZnO hexagonal c-axis, as shown by the electron diffraction image and the high-resolution TEM micrograph in Figure 2a. Au particles are found at the nanowire tips, indicating the VLS growth mode.

In order to explain the formation of both the ZnO nanowires and nanosheets, we propose a combination of the VLS and the vapor-solid (VS) growth mechanisms [1, 6]. While VLS growth occurs parallel to the growth direction of the sheet, ZnO gets deposited on the side facets via VS growth as the structure continues to grow, leading to radial growth. In the case of [0001]-oriented nanowires, this radial growth occurs with the same growth rate on each sidewall facet, because all facets have the same surface energy since they belong to the same symmetry group. The radial growth rate of these facets is low compared to the VLS growth rate along the [0001]

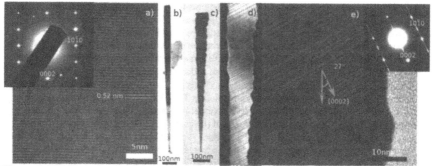

Figure 2: TEM images of a ZnO nanowire and a ZnO nanosheet grown at $T_s = 600$ °C in an ambient of 5% oxygen and 95% argon. a) High-resolution TEM image (inset shows electron diffraction pattern) and b) bright-field image of a ZnO nanowire. c) Bright-field, d) (01-10) dark-field, and e) high-resolution TEM image of a ZnO nanosheet (inset shows diffraction pattern).

direction due to the ZnO vapor impinging on the side facets of the wire either desorbing faster than chemisorbing, or being incorporated by VLS growth after diffusion along the sidewalls to the Au droplet at the tip of the nanowire.

In the case of nanosheets, the VS growth rates on the sidewall facets must be different in order for a planar, two-dimensional structure like a nanosheet to form. As discussed earlier, the growth direction of the nanosheets is not parallel to the [0001] direction. Therefore, (0001) facets are directly accessible for the impinging vapor, and VS growth can occur on the sidewalls along the [0001] direction. Since this is the reported preferred growth direction of ZnO nanostructures grown without any catalyst, we assume that nucleation on the (0001) sidewall facets is preferred over other facets during VS growth, leading to an increased growth rate along the [0001] direction relative to that of the other sidewall facets [8]. This difference in growth rates for different sidewall facets causes the structure to grow anisotropically, resulting in the planar, two-dimensional nanosheets observed in the present study.

Figure 3: TEM data of a broken nanosheet growing off a nanowire. In the center, the bright-field image of the structure is shown with arrows indicating growth directions. High-resolution images and diffraction patterns of wire and sheet regions are shown to the left and right, respectively.

On the same sample, we also observe structures that have undergone a transition from ZnO nanowire growth to nanosheet formation, as shown in Figure 3. As can be seen from the high-resolution image on the left-hand side of the figure and the inserted diffraction pattern, the upper part is a stacking fault-free single crystal with the nanowire axis parallel to the [0001] direction like the nanowires which did not exhibit a nanosheet transition observed on the same sample. As the growth continues, the growth mode suddenly changes from nanowire to nanosheet and the growth direction of the sheet no longer follows the ZnO hexagonal c-axis as denoted by the arrows in Figure 3. The high-resolution TEM image on the right-hand side of Figure 3 shows the bottom part of the structure. Numerous stacking faults similar to the ones observed for the pure nanosheets are apparent and the corresponding diffraction pattern shows the growth axis inclined to the ZnO [0001] direction. A possible explanation for this sudden shift in growth mode and growth direction could be the introduction of defects like the observed stacking faults [11], possibly due to limited surface diffusion from the substrate surface as the length of the nanowire increases. However, since perfectly crystalline nanowires of equal length are present on the sample, this cannot be the only reason. Other factors like the nanowire diameter might play an important role in the shift of growth mode. Further investigations are necessary to clarify the mechanism.

Figure 4: 45°-tilted view SEM images of ZnO nanosheets grown at a) T_s = 800 °C in a pure oxygen ambient for 60 min and b) T_s = 600 °C in a 5% oxygen and 95% argon ambient for 15 min (see also Figure 1b).

The growth of ZnO nanosheets could not only be observed at a substrate temperature of T_s = 600 °C. In fact, when using the same parameters as adopted for the ZnO nanowire growth at T_s = 800 °C, but changing the ambient gas composition from 5% oxygen – 95% argon to a pure oxygen atmosphere, ZnO nanosheets are observed on the sample, as shown in Figure 4a. Thus, the ambient gas composition, i.e. the oxygen partial pressure plays an important role in determining the morphology of these ZnO nanostructures. Contrary to the nanosheets grown at low temperature, these nanosheets exhibit smooth side surfaces with no stacking faults, as seen in the bright-field and high-resolution TEM images of Figure 5a and b, respectively. The absence of stacking faults could be explained by an increased surface diffusion due to the higher growth temperature, or possibly by annealing during post-deposition cooldown. However, the bend contours in the bright-field TEM image in Figure 5a indicate heavy strain throughout the sheet. In the inset in Figure 5a, the diffraction pattern of the ZnO nanosheet reveals an inclination of the ZnO hexagonal c-axis to the growth direction similar to the sheets grown at lower temperature.

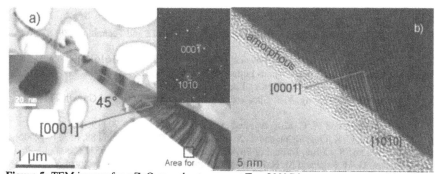

Figure 5: TEM images for a ZnO nanosheet grown at T_s = 800°C in a pure oxygen ambient for 15 min. a) Bright-field image demonstrating a growth direction inclined by 45° with respect to the ZnO [0001] direction. The insets show the electron diffraction pattern and a bright-field image of the Au particle at the tip. b) High-resolution image of single-crystalline structure.

However, for all studied nanosheets grown at $T_s = 600$ °C, the inclination angle was determined to be approximately 27°, whereas the investigated sheets grown at high temperature showed a random distribution of inclinations from 9° to 45°. From corresponding bright-field TEM images, it could also be observed that the width and the taper angle of a nanosheet were larger when its angle of inclination was increased. This supports the proposed growth mechanism for ZnO nanosheets in the sense that an increased angle of inclination involves a larger exposed surface area of (0001) sidewall facets accessible to VS growth of ZnO, thereby leading to a higher sidewall growth rate with increased inclination angle.

Another feature apparent from Figure 5b is the amorphous layer covering the nanosheet. This layer has been found for all nanostructures studied by TEM regardless of growth conditions and might have a significant impact on device performances of e.g. hybrid inorganic/organic solar cells. Further investigations on the chemical composition and on the effect of this layer on the properties of the ZnO interface might help to improve device performance.

CONCLUSIONS

In this work, ZnO nanowires and nanosheets were grown by PLD at different temperatures and oxygen partial pressures using a Au catalyst. All nanosheets were found to grow in a direction inclined to the ZnO hexagonal c-axis. Nanosheets grown at low temperature showed numerous stacking faults, while those grown at high temperature were single-crystalline. A combined VLS and VS mechanism is proposed for this growth of ZnO nanosheets.

ACKNOWLEDGMENTS

The authors would like to thank Dr. Michael Kaufman and Dr. Brian Gorman for valuable discussions. This work was supported by the Research Council of Norway under Grant No. 182092/S10 and the US National Science Foundation under Grants No. DMR-0606054 and DMR-0820518.

REFERENCES

1. G.-C. Yi, C. Wang and W. I. Park, Semiconductor Science and Technology **20**, S22-S23 (2005).
2. E. Galoppini, et al., Journal of Physical Chemistry B **110**, 16159-16161 (2006).
3. H. Ham, et al., Chemical Physics Letters **404**, 69-73 (2005).
4. Y. C. Kong, et al., Applied Physics Letters **78**, 407-409 (2001).
5. J.-H. Park, I.-S. Hwang, Y.-J. Choi and J.-G. Park, Journal of Crystal Growth **276** (1-2), 171-176 (2005).
6. R. S. Wagner and W. C. Ellis, Applied Physics Letters **4** (5), 89-90 (1964).
7. Z. W. Liu, C. K. Ong, T. Yu and Z. X. Shen, Applied Physics Letters **88**, 053110 (2006).
8. J.-J. Wu and S.-C. Liu, Journal of Physical Chemistry B **106**, 9546-9551 (2002).
9. Y. Sun, G. M. Fuge and M. N. R. Ashfold, Superlattices and Microstructures **39**, 33-40 (2006).
10. Z. W. Liu and C. K. Ong, Materials Letters **61** (16), 3329-3333 (2007).
11. Y. W. Song, S. Lee and S. Y. Lee, Journal of Crystal Growth **310** (21), 4612-4615 (2008).

Mater. Res. Soc. Symp. Proc. Vol. 1174 © 2009 Materials Research Society   1174-V09-08

Synthesis and Characterization of Flowerlike ZnO Nanoneedle Arrays on Si (100)

Boqian Yang and Peterxian Feng*
Physics Department and Institute for Functional Nanomaterials, University of Puerto Rico,
San Juan, PR 00931, U.S.A.
* Electronic mail: pfeng@cnnet.upr.edu. FAX: (787) 764 4063.

ABSTRACT

Flowerlike ZnO nanoneedle arrays have been synthesized on Si (100) substrates by pulsed laser deposition techniques. The tips of the nanoneedles are ~ 20- 50 nm in diameter and their roots are as thick as ~ 50- 100 nm. The nanoneedle arrays grow preferentially along the [0001] direction. Raman spectroscopy shows three first order optical normal modes which confirm wurtzite structure of ZnO nanoneedles. In the low frequency zone, additive modes (92, 122, 163, and 275 cm^{-1}) are observed and can be attributed to zone boundary phonons. ZnO nanoneedle arrays exhibit a strong UV luminescence emission, and two strong peaks at 3.258 eV and 3.288 eV are observed.

INTRODUCTION

One-dimensional (1D) nanostructures have attracted much attention due to their potential applications as building blocks for electronics and photonics devices [1, 2], as well as biosensors in life-science applications [3]. ZnO is a versatile material and has been used considerably for its catalytic, electrical, optical, and photochemical properties [4]. Therefore, 1D ZnO nanostructures stimulated intense attention, numerous work were carried out in this decade for the fabrication of nanostructures of various shapes i.e. (nanowires, nanobelts, nanorings, nanotubes, nanodonuts, nanopropellers, etc.) grown by different techniques such as the catalyst assisted thermal evaporation route [5], non-catalytic thermal evaporation route [6], wet chemical route [7] and solvothermal/hydrothermal [8] route, etc. Different types of substrates such as Si [9], sapphire [10], ZnO thin film coated substrate [11], metallic zinc foil [12], etc, have been reported for the self growth of various nanostructures. But due to the low cost and possibility of integration with Si-based microelectronics and microelectromechanical systems, among all the above-mentioned substrates for deposition of ZnO 1D nanostructures, Si (100) is particularly important.

In this paper, we report the preparation of 1D ZnO nanoneedle arrays without catalyst by pulsed laser deposition (PLD) techniques. The nanoneedles grow in a flowerlike assembly on the Si (100) substrate. Different tools, Scanning Electron Microscopy (SEM), Transmission Electron Microscopy (TEM), X-ray diffraction (XRD), X-ray photoelectron spectroscopy (XPS), and Raman scattering (RS) were used to investigate the morphological, structural, chemical properties of the ZnO nanostructures. The electron field emission and optical properties of ZnO nanoneedle arrays were also characterized.

EXPERIMENT DETAIL

Flower-like ZnO nanoneedles were grown on Si (100) substrates by pulse laser deposition techniques. An ArF Lambda Physik 1000 excimer laser (193 nm, ~20-30 ns, and 10 Hz repetition rate, and 200 mJ pulse energy) was used to irradiate the commercial zinc oxide target (purity up to 99.99%) at background pressure of 2.0×10^{-5} Torr in the chamber. The laser beam, focused with a 30 cm focal length fused silica lens, was incident at 45 degrees relative to the target surface. The diameter of the focused spot of the laser beam on the ZnO target was about 3 mm. The power density of the laser on the target was 1.1×10^8 W/cm² per pulse. The ZnO target was rotated at circa 200 rpm. Silicon (100) substrates were placed 4 cm away from the target. Prior to fabrication of nanostructures, the Si (100) substrates were rinsed in acetone and methanol. During deposition the substrate temperature was maintained at 600 °C whilst the deposition time was kept as ~ 30 minutes. The morphology of ZnO nanoneedle arrays was characterized using SEM and TEM, and the phase structures were tested by XRD with Cu K_α radiation. The chemical bonds of these ZnO nanostructures were examined using XPS with Mg K_α X-ray source. Room temperature micro-RS was performed using a Jobin-Yvon T64000 Triple-mate system with the radiation of 514.5 nm from a coherent argon ion laser, and a LN2 cooled charge-coupled device system was used to collect and process the scattered data. Characteristics of the field emission (I-V) from ZnO nanoneedle arrays was conducted in a custom made system where a molybdenum rod of 3 mm diameter (area: 0.071 cm2) served as the anode. The current was detected using a Keithley 6517A electrometer and electric current lower than 1×10^{-12}A was considered at the background noise level. The power supply used was a Stanford Research Systems PS350. The photoluminescence (PL) was excited using the Ti: sapphire pulsed laser (~ 4 ps) with the wavelength of 266 nm, and detected with a photonic multi-channel tube (PMT).

RESULTS AND DISCUSSION

Fig.1 (a) shows large quantities of wirelike ZnO products with an average length of 1-5 μm were formed on the surface of Si (100). These ZnO wires are arranged like many flowers with a uniform diameter of 4 - 8 μm, due to the (100) orientation of Si substrates. A higher magnification SEM image Fig.1 (b) reveals that the ZnO products have the shape of a needle with a sharp tip and thick root.

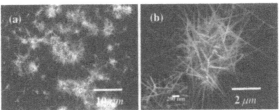

Fig. 1. Low magnification (a) and higher magnification (b) SEM images of flowerlike ZnO nanoneedles on the Si (100) substrate

Fig.2 shows XRD patterns with highly (002) preferential orientation of the nanoneedles on Si (100). The (100), (002), and (101) diffraction peaks were identified, which correspond to the hexagonal ZnO structure. The lattice parameter 'c' and crystallite grain size of the samples have been estimated as 0.5180 nm and 28.0 nm respectively, from the XRD patterns.

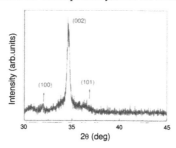

Fig. 2. XRD pattern of ZnO nanoneedles prepared on Si(100) substrate

The TEM image of a single nanoneedle shown in Fig.3 (a), reveals the bead-chain of nano-grains with diameter of 20 -30 nm along the nanoneedle, which indicates the vapor-solid continuous deposition process [13] for growth of nanoneedle arrays on Si (100). Fig. 3 (b) and (c) shows the node-structure of nanoneedle with a sharp tip of 10- 15 nm and thick root of 30- 50 nm in diameter. Here, the size of the nano-grains is in agreement with the value estimated from XRD patterns. Fig. 3 (d) is a selected-area electron diffraction (SAED) pattern taken from the root part of a single ZnO nanoneedle, which proves that the ZnO nanoneedle is highly (002) preferential crystalline.

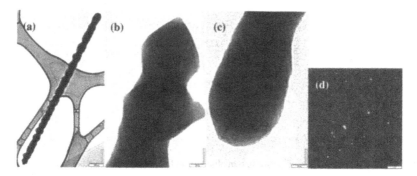

Fig. 3. TEM images of (a) the single nanoneedle with scale bar of 200 nm, (b) the tip of nanoneedle with scale of 20 nm, (c) the root of nanoneedle with scale of 20 nm, and (d) a SAED pattern from the root

The composition of the ZnO nanoneedle was analyzed using XPS techniques, which provides information on the oxidation state of each component as well as the composition of the sample surface. Fig.4 (a) shows the full XPS spectrum of the sample, and no peaks for any other elements except Zn, O and C are observed. The peak of C1s came from the carbon tape used during the measurement. High-resolution XPS spectra of Zn 2p3/2 and O 1s states are shown in Fig.4 (b) and (c). The binding energy of Zn 2p3/2 is around 1022.9 eV, which is larger than the value of Zn in bulk ZnO. No metallic Zn with a binding energy of 1021.50 eV is observed, which confirms that Zn exists only in the oxidized state [14]. The O 1s state peak of ZnO nanoneedles split into two peaks as shown in Fig.5 (c). The higher binding energy component located at 532.8 ± 0.2 eV is usually attributed to the presence of some Zn $(OH)_2$ phase in the samples which may be formed by absorbing moisture from free atmosphere [12]. The peak at 531.2 ± 0.2 eV can be due to the O-Zn bond formation, which is attributed to O^{2-} ions on wurtzite structure of hexagonal Zn^{2+} ion array [15].

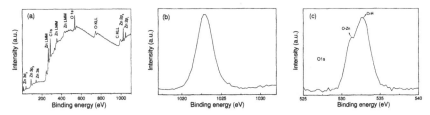

Fig. 4. XPS spectra of synthesized ZnO nanoneedles: (a) full range survey spectrum; (b) Zn 2p3/2 spectrum; and (c) O 1s spectrum

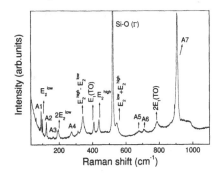

Fig. 5. Raman scattering spectra of ZnO nanoneedles prepared on Si (100)

To study the nature of crystalline quality of ZnO nanoneedles, we have carried out micro-Raman measurements as shown in Fig.5. Three optical normal modes locate at 101, 342, 407, and 438 cm^{-1} corresponding to E_2^{low}, $E_1(TO)$, and E_2^{high}, respectively [16]. Along with these normal modes, second order modes at 195, 342, 544 and 790 cm^{-1} are observed, which are attributed to $2E_2^{low}$, $E_2^{high} - E_2^{low}$, $E_2^{high} + E_2^{low}$, and $2E_1(TO)$. Except these normal first and second order modes, seven additive peaks A1-A7 were observed. Peaks assigned to A5, A6, and A7 are probably due to the multiphonon processes or defect in ZnO nanoneedles [16]. The peaks in low frequency region at 92, 122, 163, and 275 cm^{-1} can not be associated with any first or second

66

order mode at the center of the Brillouin zone. These additive modes A1-A4 can be explained by the activation of Zone boundary phonons, which can be activated either by dopant incorporation, nanosized growths, or anything that can break crystal translational symmetry [17]. Comparing to the Raman spectra from ZnO thin films [18], the nanosized growth of nanostructures excites more Raman modes as shown in Fig. 5.

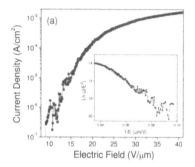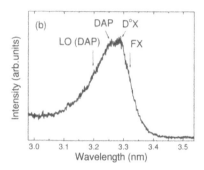

Fig. 6. (a) Emission current density (the inset reveals that the field emission follows FN behavior), and (b) Room temperature PL spectra, from ZnO nanoneedle arrays on Si (100)

Fig.6 (a) shows the field emission property from ZnO nanoneedles at a 2×10^{-6} Torr background pressure at room temperature. The turn-on electric field is about $9.0\,V\mu m^{-1}$ at current density of $0.1\,\mu Acm^{-2}$ for ZnO nanoneedles. The emission current density reaches about $1\,mAcm^{-2}$ at an applied field of about $30.0\ V\mu m^{-1}$ for nanoneedles. The Fowler-Nordheim (FN) plots shown in the inset of Fig.6 (a) exhibit rough-linear behavior in a measurement range, and thus the emission current is mainly caused by the quantum tunnel effect.

The optical property of ZnO nanoneedles on Si (100) was characterized with the room temperature PL spectra in UV region as shown in Fig.6 (b). The UV peaks can be considered a combination of multiple peaks due to donor bound-exciton (D^oX), donor-acceptor-pair (DAP), and free exciton (FX) transitions. Two strong PL peaks at 3.258 eV and 3.288 eV are observed, which are probably attributed to donor-acceptor-pair (DAP) and donor-bound-exciton (D^oX) transitions, respectively [19]. The broader shoulder on the lower energy side is due to the multiple longitudinal optical-phonons (LO) replicas of the DAP band, and the high-energy shoulder is assigned to free exciton (FX) transition. The ZnO nanoneedles can be used as UV light-emitting materials.

CONCLUSIONS

In summary, flowerlike 1D ZnO nanoneedle arrays have been synthesized via vapor-solid processing on Si (100) using the PLD techniques. The TEM analysis reveals the bead-chain of nano-grains with diameter of 20 -30 nm along a single nanoneedle. XPS confirms the chemical purity of the flowerlike ZnO nanoneedle arrays. Raman measurement shows the wurtzite structure of ZnO nanoneedles, and the growth of nanoneedles breaks crystal translational symmetry of ZnO. The emission current density reaches about 1 $mAcm^{-2}$ at an applied field of about 30.0 $V\mu m^{-1}$, and the dominant peaks of UV emission are attributed to D^oX and DAP transitions.

ACKNOWLEDGMENTS

This work is partially supported by NSF-EPSCoR fellowship, NSF-DMR (0706147), DoD (W911NF-07-1-0014). We would like to thank Mr. William for assistance of Raman measurements, Mr. Josue Morales for SEM measurements, Prof. Gerardo Morell for field emission measurements, and Prof. Marc Achermann and F. Madalina for the help in the photo-luminescence measurements.

REFERENCES

1. D. Appell, *Nature (London)* **419**, 553 (2002)
2. X.F. Duan, Y. Huang, Y. Cui, J.F. Wang, and C. M. Lieber, *Nature (London)* **409**, 66 (2001)
3. Y. Cui, Q. Wei, H. Park, and C.M. Lieber, *Science* **293**, 1289 (2001)
4. L. Vayssieres, K. Keis, A. Hagfeldt, and S.-E. Lindquist, *Chem. Mater.* **13**, 4386 (2001)
5. Huang M H, Mao S, Feick H, Yan H Q, Wu Y Y, Kin H, Weber E, Russo R and Yang P D, *Science* **292**, 1897 (2001)
6. Kar S, Pal B N, Chaudhuri S and Chakravorty D, *J. Phys. Chem. B* **110**, 4605 (2006)
7. Tian Z R R, Voigt J A, Liu J, Mckenzie B, McDermott M J, Rodriguez M A, Konishi H and Xu H F, *Nat. Mater.* **2**, 821 (2003)
8. Dev A, Kar S, Chakrabarti S and Chaudhuri S, *Nanotechnology* **17**, 1533 (2006)
9. Ye Sun, Katherine E Addison and Michael N R Ashfold, *Nanotechnology* **18**, 495601 (2007)
10. M.Lorenz, E.M. Kaidashev, A. Rahm, Th. Nobis, J. Lenzner, G. Wagner, D. Spemann, H. Hochmuth, M. Grundmann, *Appl. Phys. Lett.* **86**, 143113 (2005)
11. Jiansheng Jie, Guanzhong Wang, Yiming Chen, Xinhai Han, Qingtao Wang, Bo Xu, and J.G. Hou, *Appl. Phys. Lett.* **86**, 031909 (2005)
12. Tandra Ghoshal, Subhajit Biswas, Soumitra Kar, Apurba Dev, Supriya Chakrabarti and Subhadra Chaudhuri, *Nanotechnology* **19**, 065606 (2008)
13. Pan Z.W., Dai Z.R., Wang Z.L., *Science* **291**, 1947 (2001)
14. M.N. Islam, T.B. Gosh, K.L. Chopra, H.N. Acharya, *Thin Solid Films* **280**, 20 (1996)
15. Zheng-Bin Gu, Ming-Hui Lu, Jing Wang, Di Wu, Shan-Tao Zhang, et.al, *Appl. Phys. Lett.* **88**, 082111 (2006)
16. T.C. Damen, S.P.S. Porto, and B. Tell, *Phys. Rev.* **142**, 570 (1966); J.M. Calleja and M. Cardona, *Phys. Rev. B* **16**, 3753 (1977)
17. Harish Kumar Yadav, K. Sreenivas, Vinay Gupta, and R.S. Katiyar, *Journal of Applied Physics* **104**, 053507 (2008)
18. Boqian Yang, Ashok Kumar, Peter Feng, and R.S. Katiyar, *Appl. Phys. Lett.* **92**, 233112 (2008)
19. U. Ozgur, and H. Morkoc, *Optical properties of ZnO and related alloys, Chapter 5, Zinc Oxide Bulk, Thin Films and Nanostructures*, edited by chennupati Jagadish and Stephen J. Pearton, Elsevier, 2006, pp175

Mater. Res. Soc. Symp. Proc. Vol. 1174 © 2009 Materials Research Society 1174-V09-05

Investigation of Characteristics of Multi-Function ZnO Thin Film Deposited With Various Argon and Oxygen Ratios

Che-Wei Hsu[1], Tsung-Chieh Cheng[2], Chun-Hui Yang[3], Yi-Ling Shen[3], Jong-Shin Wu[1*], Sheng-Yao Wu[2], Wen-Hsien Huang[3]

[1]Department of Mechanical Engineering, National Chiao Tung University, 1001 Ta-Hsueh Road, Hsinchu 30050, Taiwan

[2] Department of Mechanical Engineering, National Kaohsiung University of Applied Science, 415 Chien Kung Road, Kaohsiung 807, Taiwan

[3] National Nano Device Laboratories, No. 26, Prosperity Road I, Science-based Industrial Park, Hsinchu 30078, Taiwan

ABSTRACT

The ZnO thin film was deposited on a glass substrate at RT by the RF reactive magnetron sputtering method. Structural, chemical, optical, and hydrophilic/hydrophobic properties are measured by using a surface profilometer, an x-ray diffractometry (XRD), an x-ray photoelectron spectroscopy (XPS), a UV-VIS spectrophotometer, and a contact angle system, respectively. Results show that the deposition rate decreases with increasing $O_2/(Ar+O_2)$ ratio. In additiion, the best stoichiometric and quality of ZnO thin film was observed at 0.30 of $O_2/(Ar+O_2)$ ratio, which shows the smallest FWHM and the strongest O-Zn strength. Regardless of $O_2/(Ar+O_2)$ ratio effect or thickness effect, high transmittance (> 86%) in the visible region is observed, while the UV-shielding characteristics depend upon both the magnitude of film thickness. The film thickness plays a more prominent role in controlling optical properties, especially in the UV-shielding characteristics, than the $O_2/(Ar+O_2)$ ratio. However, the hydrophobic characteristics can be obtained when the glass coating with ZnO thin films. In general, with properly coated ZnO thin film, we can obtain a glass substrate which is highly transparent in the visible region, has good UV-shielding characteristics, and possesses highly hydrophobic characteristics (self-clean capability), which is highly suitable for applications in the glass industries.

INTRODUCTION

In recent years, ZnO is one of the promising candidate materials which has been extensively investigated for various applications in many fields, such as optoelectronics [1-2], piezoelectric [3], and energy & environment [4-5], to name a few. Wide applications of ZnO thin film arise from several unique material properties. Therefore, ZnO is considered as one of the most interesting semiconductors of II–VI compounds. ZnO films have been grown by various deposition methods, such as sputtering [6-7], chemical bath deposition (CBD), spray pyrolysis, pulsed laser deposition (PLD), and metal organic chemical vapor deposition (MOCVD). However, understanding of the relationship between thin-film properties and plasma properties should greatly help to produce high-quality ZnO thin film in sputtering technique. Overdosed ultraviolet (UV) irradiation has become a serious problem due to ozone depletion globally and causes damaging effects on human's health. Hence, how to prevent the UV light from direct contact with human while allow most of the visible light to fill in the living space is a critical issue in modern glass industry or human life [8-10]. In addition, highly hydrophobic glass surface is also strongly necessary, which can greatly reduce the cost of maintenance. In the present study, we focus on investigating the effects of $O_2/(Ar+O_2)$ ratio and film thickness on the multifunction glasses and films. In addition, relation between deposition rates, plasma conditions and general physical properties of the ZnO thin films are also addressed in the paper.

EXPERIMENT

In the present study, the ZnO thin film was deposited on glass substrate at room temperature (RT) with a Zn target by a RF reactive magnetron sputtering. Argon and oxygen are used as the working (discharge) and reactive gas in mass flow rate, respectively. Distance between the target and substrate is kept as 8 cm unless otherwise specified. ZnO thin films were grown using a fixed RF power of 100 W at a constant working pressure of 15 mtorr. Two major test conditions for preparing ZnO thin film include: 1) $O_2/(Ar+O_2)$ ratio ranges from 0.1 to 1.0 with a constant film thickness of 60 nm; 2) Deposition time ranges from 5 to 90 min. with a fixed $O_2/(Ar+O_2)$ ratio of 0.25. Film thickness was measured by a surface profilometer. Crystal structure and orientation of the films was determined by the XRD. Composition and chemical state of oxygen in ZnO thin films was studied through an XPS using AlKα X-ray source (Model VG Scientific Microlab 310F). All obtained spectra were calibrated to a C $1s$ electron peak at 284.6 eV. Optical transmittance of the ZnO films was investigated using a UV-VIS spectrophotometer. Contact angle was observed by a contact angle system with a universal surface tester.

RESULTS AND DISCUSSIONS

In general, the deposition rate decreases monotonically with increasing $O_2/(Ar+O_2)$ ratio [11] as shown in Fig. 1, which is strongly correlated with the concentration of argon and oxygen molecules. Strictly speaking, the ionization energy of oxygen (48.76 eV) is higher than argon (15.76 eV) [12]. For this reason, either the ion or electron number density decrease with increasing $O_2/(Ar+O_2)$ ratio [13-14]. Hence, the oxygen molecules absorbed more energy to ionize and result in decreasingly ionization probability with increasing $O_2/(Ar+O_2)$ ratio. This will decrease the ion flux and sputtering yield [11, 14], which results in the lower deposition rate at higher $O_2/(Ar+O_2)$ ratio.

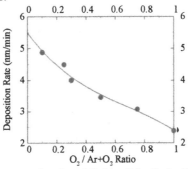

Figure 1. The deposition rate as a function of $O_2/(Ar+O_2)$ ratio for ZnO thin film deposition.

In this study, all of these ZnO thin films are in poly-crystalline with a hexagonal structure and the major orientation is (0002). Fig. 2a exhibits the thickness effect on the ZnO thin films. Results clearly show that the ZnO thin films grown less than 5 min. (smaller than 50 nm in thickness) are almost in amorphous-like phase [15] because the major orientation of ZnO (0002) is not obvious. As the ZnO thin film grows over 10 min. (large than 50 nm in thickness), crystalline structure of orientation (0002) begins to appear and become dominant in the film structure with further increasing of deposition time [7]. In addition, ZnO thin films with enough film thickness (60 nm) are dominated by the (0002) orientation no matter what the $O_2/(Ar+O_2)$ ratio is, but the peak intensities remain approximately the same at varying $O_2/(Ar+O_2)$ ratios (not shown in the paper). In order to understand the slight influence of $O_2/(Ar+O_2)$ ratio in the ZnO thin film structural, in which the FWHM of ZnO (0002) and the grain size was calculated by Schererr equation [7] as shown in Fig. 2b. Further, the best film of ZnO thin film was able to decide that the minimum value of FWHM has been obtained at 0.3 $O_2/(Ar+O_2)$ ratio in this study, which is similar to that obtained in [14]. Furthermore, grain size increases from 29.28 nm at 0.1 of $O_2/(Ar+O_2)$ ratio up to 35.68 nm at 0.3 of $O_2/(Ar+O_2)$ ratio, and then decreases gradually down to 30.86 nm eventually at 1.0 of $O_2/(Ar+O_2)$ ratio. Finally, the maximum grain size also occurs at 0.3 of $O_2/(Ar+O_2)$ ratio that consistent with best film (the lowest FWHM) of ZnO film.

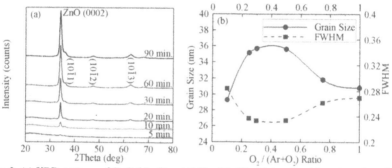

Figure 2. (a) XRD patterns of ZnO thin film at 0.25 of $O_2/(Ar+O_2)$ ratio with various deposition time and (b) the FWHM of ZnO (0002) and the calculated grain size as a function of $O_2/(Ar+O_2)$ ratio.

In summary, the binding energy of Zn 2p3 peak is about 1021.8 eV [16] for all gas ratios. We also observed no obvious change in its chemical composition in this study. Nevertheless, the chemical state of O 1s peak has appreciable shift as shown in Fig. 3a. In brief, the binding energy of O 1s peaks were separately into two components, namely O^I (the lower binding energy at 529.8 eV) [17] and O^{II}(the lower binding energy at 531.6 eV) [18], respectively. The O^I peak is attributed to O^{2-} ions on wurtzite ZnO thin films (O-Zn bond); but the O^{II} peak could be caused by the loosely bound oxygen from absorbed H_2O or by the defect O^{2-} ions in the surface during film growth process. In order to understand the importance of the O^I peak (O-Zn bond) in ZnO characteristics; we try to analyze the degree of O^I peak (O-Zn bond) in O 1s as shown in Fig. 3b. Maximum value of both O^I/O_{total} and O^I/O^{II} ratios occurs at approximately 0.3 of $O_2/(Ar+O_2)$

ratio. This is also consistent with the smallest FWHM and largest grain size observed by XRD analysis which occurs at the same gas ratio. Therefore, we deduced that the content of O^I peak (O-Zn bond) is important in deciding ZnO film properties. The above observation indicates that the higher the O-Zn bond ratio the better the ZnO film quality, which is also consistent with the XRD trend.

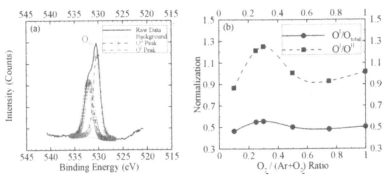

Figure 3. (a) The XPS spectra of O 1s peak of ZnO thin film at 0.3 of $O_2/(Ar+O_2)$ ratio and (b) the content ratio of O^I/O_{total} and O^I/O^{II}, respectively.

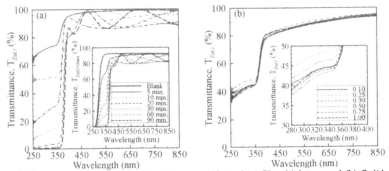

Figure 4. The transmittance of ZnO thin film with effect of (a) film thickness and (b) $O_2/(Ar+O_2)$ ratio. Inset of figure 4a and 4b are showing the transmittance of ZnO thin film on glass substrate ($T_{ZnO/Glass}$%) and the transmittance of ZnO thin film (T_{ZnO}%) in the UV region, respectively.

Furthermore, Fig. 4a shows the higher transmittance in the range of 400-700 nm from 98.3% to 87.7% with various deposition times (thickness effect). Correspondingly, the UV-shielding characteristic decreases from 78.3% down to 13.1% with increasing deposition time. Moreover, the inset of Fig. 4a presents the ZnO thin film on glass still has high transparent (80-90%) and better UV-shielding characteristics with various deposition times. Secondly, Fig. 4b is observed that all of $O_2/(Ar+O_2)$ ratio cases exhibit higher transparent (over 86.5%) and partly UV-shielding characteristics (from 55% to 49%). Regardless of pure ZnO thin film or ZnO on glass, results show that the transmittance spectra difference of various film thicknesses have obvious variation, especially in UV region (250-400 nm). So, we concluded that the film thickness plays a key role in optical characteristics than the $O_2/(Ar+O_2)$ ratio. We expect that the optimum thick film thickness will produce the completely UV-shielding with highly transparent ZnO thin film in some applications. Incidentally, in terms of color theory, the color characteristics can also be observed form these transmittance spectra by additive mixing of color-mixing [19]. For example, according to Fig. 4b, due to the relatively strong long wavelength and the relatively weak short wavelength in visible region couple with together to display the light yellow characteristics in all approximation 60 nm ZnO thin films. As mentioned above, the various color characteristics have been listed in Table 1.

Table 1. The various thickness and color characteristics for all ZnO thin films

Specimens Name	Thickness	Color Characteristics
5 min.	~ 43.4 nm	Snow
10 min.	~ 51.4 nm	Floral White
20 min.	~ 73.4 nm	Khaki1
30 min.	~ 129.6 nm	Pale Violet Red1
60 min.	~ 235.6 nm	Peach Puff1
90 min.	~ 369.9 nm	Sea Green2
$O_2/(Ar+O_2)$ Ratios (0.1~1.0)	~ 60 nm ± 1.1%	Light Yellow

In Fig. 5a, the blank glass displays hydrophilic characteristics with the value of contact angle about 55.8°, while in Fig. 5b and 5c, the glass substrate coated with ZnO thin film show very large contact angle over 92.7°. Note for all the tested cases all contact angles are at least 30° larger than the case of blank glass substrate. The maximum value of contact angle of ZnO on glass is about 111.3°. The result clearly shows that the ZnO thin film is able to effectively modify the surface of glass substrate from hydrophilic to hydrophobic.

Figure 5. Hydrophobic characteristics of the glass substrate (a) without ZnO thin film (b) with ZnO thin film at 5 min. (c) with ZnO thin film at 0.25 of $O_2/(Ar+O_2)$ ratio.

CONCLUSIONS

In the current study, the ZnO thin film was successfully deposited on a glass substrate at RT by a RF reactive magnetron sputtering method. Results show that the deposition rate decreases with increasing $O_2/(Ar+O_2)$ ratio. The mainly reason is the ionization energy of oxygen (48.76 eV) higher than argon (15.76 eV) and it will decrease the ionization probability and sputtering yield, which results in the lower deposition rate at higher $O_2/(Ar+O_2)$ ratio. In additionally, the optimum stoichiometric (the strongest O-Zn bond ratio) and quality (the lowest FWHM) of ZnO thin film was observed at 0.3 of $O_2/(Ar+O_2)$ ratio by XRD and XPS. However, the lack oxygen concentration will produce more crystallographic faults; but the excess oxygen concentration will destroy the stoichiometric. Regardless of $O_2/(Ar+O_2)$ ratio effect or thickness effect, high transmittance (> 86%) in the visible region is observed, while the UV-shielding characteristics depend upon both the magnitude of film thickness. The film thickness plays a more prominent role in controlling optical properties, especially in the UV-shielding characteristics, than the $O_2/(Ar+O_2)$ ratio. However, the hydrophobic characteristics can be obtained when the glass coating with ZnO thin films.

ACKNOWLEDGMENTS

The authors would like to express their sincere thanks to the financial support of National Science Council of Taiwan through NSC 96-2628-E-009-134-MY3 and NSC 96-2628-E-009-136-MY3. Also the instrumentation provided by National Nano Device Laboratory of Taiwan is also highly appreciated.

REFERENCES

1. S. Tüzemen and Emre Gür, Opt. Mater. 30 (2007) 292.
2. John F. Wager, Science 300 (2003) 1245.
3. Thomas Thundat, Nature Nanotechnol. 3 (2008) 133.
4. Y. U. Feng, Journal of Wuhan University of Technology-Mater. Sci. Ed., 22 (2007) 385.
5. E. Ando and M. Miyazaki, Thin Solid Films 516 (2008) 4574.
6. Kun Ho Kim, Ki Cheol Park and Dae Young Ma, J. Appl. Phys. 81 (1997) 7764.
7. Yuantao Zhang, Guotong Du, Dali Liu, Xinqiang Wang, Yan Ma, Jinzhong Wang, Jingzhi Yin, Xiaotian Yang, Xiaoke Hou, and Shuren Yang, J. Cryst. Growth 243 (2002) 439.
8. Marcelo N. Ayala, Ralpb Micbael, and Per G. Söderberg, Invest. Ophthalmol. Vis. Sci. 41 (2000) 3539.
9. Shinryo Yabe, Mika Yamashita, Shigeyoshi Momose, Kazuyuki Tahira, Sakae Yoshida, Ruixing Li, Shu Yin, Tsugio Sato, Int. J. Inorg. Mater. 3 (2001) 1003.
10. Ruixing Li, Shinryo Yabe, Mika Yamashita, Shigeyosi Momose, Sakae Yoshida, Shu Yin, Tsugio Sato, Mater. Chem. Phys. 75 (2002) 39.
11. Rashmi Menon, K. Sreenivas, and Vinay Gupta, J. Appl. Phys. 103 (2008) 094903.
12. Charles Kittel, Introduction to Solid State Physics, 8th Edition, Wiley, 2004, pp 54.
13. T. Nagata, A. Ashida, N. Fujimura, and T. Ito, J. Appl. Phys. 95 (2004) 3923.
14. Yee-Shin Chang, Jyh-Ming Ting, Thin Solid Films 398-399 (2001) 29.
15. Y. Yoshino, K. Inoue, M. Takeuchi, K. Ohwada, Vacuum 51 (1998) 601.
16. Li-Wen Lai, Chinf-Ting Lee, Mater. Chem. Phys. 110 (2008) 393.
17. M. Chen, X. Wang, Y. H. Yu, Z. L. Pei, X. D. Bai, C. Sun, R. F. Huang, L. S. Wen, Appl. Surf. Sci. 158 (2000) 134.
18. R.E. Marotti, C.D. Bojorge, E. Broitman, H.R. Cánepa, J.A. Badán, E.A. Dalchiele, A.J. Gellman, Thin Solid Films, 517 (2008) 1077.
19. Roy S. Berns, Principle of Color Technology, 3rd Edition, John Wiley & Sons Inc., 2002, Ch 6.

Mater. Res. Soc. Symp. Proc. Vol. 1174 © 2009 Materials Research Society   1174-V02-08

Surfactant-Mediated Synthesis of Functional Metal Oxide Nanostructures Via Microwave Irradiation-Assisted Chemical Synthesis

Sanjaya Brahma[1], S. A. Shivashankar[1]
Materials Research Centre, Indian Institute of Science, Bangalore-560012, India

ABSTRACT

In the present work we report a rapid microwave irradiation-assisted chemical synthesis technique for the growth of nanoparticles, nanorods, and nanotubes of a variety of metal oxides in the presence of an appropriate surfactant (cationic, anionic, non ionic and polymeric), without the use of any templates. The method is simple, inexpensive, and helps one to prepare nanostructures in quick time, measured in seconds and minutes. This method has been applied successfully to synthesize nanostructures of a variety of binary and ternary metal oxides such as ZnO, CdO, Fe_2O_3, CuO, Ga_2O_3, Gd_2O_3, $ZnFe_2O_4$, etc. There is an observed variation in the morphology of the nanostructures with changes in different process parameters, such as microwave power, irradiation time, identity of solvent, type of surfactant, and its concentration.

INTRODUCTION

With the advent of technologically sophisticated facilities, the synthesis strategies of metal oxide nanostructures, with potential application in many fields [1], have either been changed, or modified, and novel synthesis procedures undertaken by researchers. Recently developed synthesis techniques include the sol-gel technique [2], hydrothermal synthesis [3], solvothermal process [4], chemical co-precipitation [5], polyol synthesis, thermal decomposition [6] and spray pyrolysis [7]. While there has been a substantial improvement in these techniques, they are often energy-intensive and time-consuming. Thus, there has been a need to modify the existing synthesis techniques or to design a novel synthesis technique /procedure which consumes comparatively less time and energy, is less cumbersome, while requiring no expensive equipment. Microwave irradiation-assisted chemical synthesis [8-10] is such a process, and can be carried out in a simple "oven" at power varying from 160 – 800 W. In the present work, we have tried to explore the capability of this process in preparing different nanostructures of ZnO in particular and nanoparticles of other oxide materials such as CdO, Fe_2O_3, Ga_2O_3, Gd_2O_3, using the corresponding metal acetylacetonate complex, in presence of a surfactant. The as-prepared oxide materials either need a brief exposure to heat at a modest temperature to remove the surfactants or, no treatment at all, because they are already pure and crystalline. There is a discernible variation in the morphology of the nanostructures when different process parameters are changed, such as microwave power, irradiation time, identity of solvent, type of surfactant used and its concentration. Cationic, anionic, nonionic and polymeric surfactants have been used to generate a variety of nanostructures.

EXPERIMENTAL

Metal oxide nanoparticles and nanostructures were prepared from the corresponding metal acetylacetonate, which is usually a crystalline solid at room temperature. The schematic molecular structure of such a complex is given in figure 1, showing the direct metal-to-oxygen bonds present in the complex, making it a suitable precursor for the synthesis of oxides. In the present procedure, one gram of the precursor material is taken in a round-bottomed flask and dissolved in appropriate amount of solvent (most commonly ethanol or methanol) and stirred for 15 minutes. A solution of the surfactant polyvinyl pyrrolidone (PVP, molecular weight ~360,000) in double-distilled water is prepared (0.3 g PVP in 40 ml water), and added to the previously prepared metal acetylacetonate solution, followed by 15 minutes of stirring. (The surfactant was added to prevent agglomeration of particles that result from microwave irradiation.) The solution so obtained in the round-bottomed flask was placed in a domestic-type microwave oven (2.45 GHz, power variable from 160 – 800 W). Both the composition of the solutions prepared and the parameters of the microwave irradiation were varied. It may be noted that the microwave synthesis system was equipped with a water-cooled condenser (reflux system) placed outside the microwave oven.

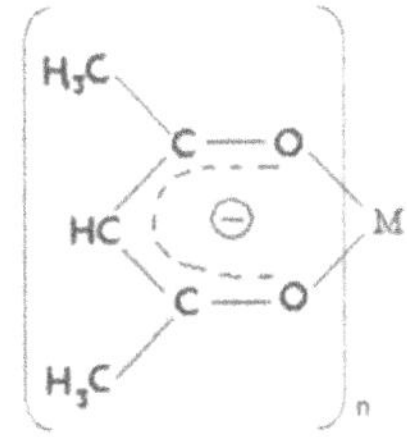

Figure 1. Schematic molecular structure of the acetylacetonate complex of a metal M

Microwave irradiation of the solution was found to result in a colloidal suspension. Centrifugation of this suspension, followed by washing with acetone and distilled water, yielded a fine powder, which was heated for a few minutes in air at 450°C to remove the residual surfactant. All the samples were characterized by powder XRD (Philips diffractometer, Model 3710 MPD, Cu-K$_\alpha$ radiation). XRD patterns were recorded from 20° to 80° (2θ) with a scanning step of one degree per minute. The size distribution and morphology of the particles in the samples were analyzed by field emission scanning electron microscopy (FE-SEM SIRION XL-40) and transmission electron microscopy (TEM-TECNAI F-30). Selected area electron diffraction (SAED) was carried out in the TEM to analyze the structure of individual fine particles in the samples. Elemental analysis was carried out by an energy-dispersive X-ray analyzer attached to the SEM and the TEM.

DISCUSSION

ZnO nanostructures have been synthesized from the Zn acetylacetonate ($C_{10}H_{14}O_4$ .Zn) complex commonly denoted by Zn (acac)$_2$. The XRD patterns of the as-prepared powder samples prepared with microwave irradiation for 30 s, 1 min, and 5 min are shown in figure 2a, which clearly indicates that each sample is well-crystallized. All the diffraction peaks can be indexed to zinc oxide of the hexagonal würtzite structure. The XRD data yield lattice constants of a=3.2478 Å and c =5.2035 Å [11]. The absence of diffraction peaks from impurities or unreacted precursors confirms the purity of the hexagonal ZnO phase formed. The general morphology of the as-prepared ZnO product (800 W, 30 seconds) is shown in the scanning electron micrographs of figure 2, of which figure 2b reveals that the sample obtained in 30 seconds is composed of very fine ZnO nanoparticles of ~20 nm. The particle size and shape vary with the variation of process parameters, such as microwave power, irradiation time, and concentration of the surfactant (PVP). It is observed that, keeping the microwave power and surfactant concentration fixed, the variation in irradiation time yields different nanostructures of ZnO, *i.e.,* when the microwave irradiation time approaches 5 min, the resulting mixture gives ZnO nanorods having very distinct hexagonal and nearly equal cross section, and nanotubes having wall thickness ~40 nm and inner diameter in the range of 140-160 nm., as depicted in figure 2c and figure 2d. The distinctive feature of the tubes is that they possess smooth and regular hexagonal cross sections, although no conventional template has been used in their preparation. These nanorods and nanotubes are highly crystalline and individual nanorods are single crystals, as shown by the SAED pattern (figures 2e and 2f). Each pattern consists of very distinct bright spots corresponding to the different crystallographic planes of würtzite ZnO. These spots have been indexed and the zone axis has been determined, as shown in the electron diffraction pattern. In the SAED pattern of the ZnO nanotube, each individual spot is associated with an additional spot, which is attributed to the shape of the nanotube.

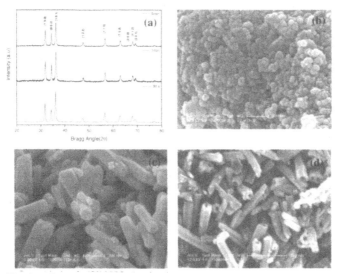

Figure 2. (Continued)

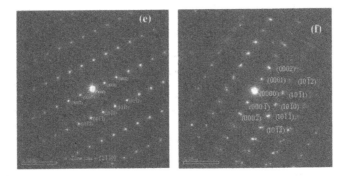

Figure 2. ZnO nanostructures synthesized with polyvinyl pyrrolidone (PVP) as a surfactant. (a) X – ray diffraction pattern of ZnO nanoparticles synthesized in 30 s, 1 min, and 5 min of microwave irradiation. (b) SEM image of ZnO nanoparticles. (c) & (d) SEM image of ZnO nanorods and nanotubes synthesized with 5 mins of irradiation. (e) & (f) SAED pattern of ZnO nanorods and nanotubes, respectively.

The powder material obtained by the above process using PVP as surfactant requires a thermal treatment for ~30 minutes in air at 450 °C to remove the residual polymer completely. It is, therefore, desirable to use a surfactant that obviates the need for any thermal treatment, *i.e.,* the as-prepared powder material should be pure and crystalline. To achieve synthesis of ZnO nanostructures in this manner, we have tried to vary the surfactant as well as the solvent. Specifically, we have employed ionic surfactants such as sodium dodecyl sulfate (anionic surfactant, commonly called SDS), hexadecyl trimethyl ammonium bromide (CTAB), which is a cationic surfactant, as well as a non-ionic surfactant called Triton X 100. All these surfactants are found to be effective in the synthesis of oxide nanostructures by the present method. Similarly, we have changed the solvent from ethanol to a long chain alcohol *i.e.,* decanol, and found that it works as both the solvent and the surfactant. The powder materials obtained using the above surfactants and solvent are very uniform in size and shape, pure, and crystalline.

Using an anionic surfactant such as sodium dodecyl sulphate, ZnO "nano-bipods" have been obtained, as shown in figure 3a, while figure 3b depicts straight ZnO nanorods which have been synthesized using the cationic surfactant CTAB. A very different nanostructure is formed, *i.e.,* ZnO nanoflowers as shown in figure 3c, when the nonionic surfactant Triton X 100 is employed. A fascinating morphology of ZnO was obtained when decanol was used as a solvent, which also works as the surfactant. The structure is that of a ZnO nanosphere (~120 nm) embedded by nanoparticles having average diameter ~ 15 nm. Variations in the morphology of the ZnO nanostructures with the variation of the concentration of decanol, with the mixing of decanol with a primary alcohol such as ethanol or with some other long chain alcohol like hexanol, have been observed, and are under further investigation.

Figure 3. (a) ZnO "nano-bipods" with sodium dodecyl sulphate as anionic surfactant (b) ZnO nanorods with CTAB as cationic surfactant (c) ZnO nano flowers with Triton X 100 as nonionic surfactant and d) ZnO nanosphere with decanol as solvent.

The microwave irradiation process has been applied to obtain nanoparticles of other metal oxides, such as CdO, Fe_2O_3, Ga_2O_3 and Gd_2O_3, as shown in figure 4. Nanoparticles of CdO synthesized using Cd $(acac)_2$ as the precursor material are shown in figure 4a, while Fe_2O_3 nanoparticles prepared using $Fe(acac)_3$ as the starting material are shown in figure 4b. Similarly, Ga_2O_3 and Gd_2O_3 nanoparticles as shown in figure 4c and figure 4d have been prepared by using the corresponding acetylacetonate precursors. The synthesis of these oxides requires conditions comparable to those used for the synthesis of ZnO, though the precise details vary. Preliminary work shows that nanoparticles of binary oxides such as CuO and MnO, and of ternary oxides such as $ZnMn_2O_4$, may also be synthesized in the same manner.

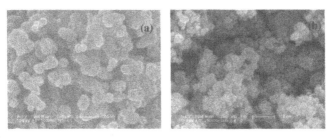

Figure 4. (Continued)

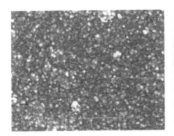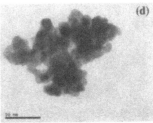

Figure 4. a) Nanoparticles of Cadmium oxide (CdO) b) Iron oxide (Fe$_2$O$_3$), c) Gallium oxide (Ga$_2$O$_3$) d) Gadolinium oxide (Gd$_2$O$_3$).

CONCLUSIONS

Microwave irradiation-assisted chemical synthesis technique has great potential in the synthesis of nanoparticles and nanostructures of a variety of different oxide materials, within a very short time, i.e., a few seconds or minutes. The as-prepared powder materials are pure and crystalline, and require no further processing, if an appropriate surfactant is employed.. The shape and size of the nanocrystallites can be tailored by employing different process parameters.

REFERENCES

1. J. A. Rodriguez and M. Fernandez-Garcia, Synthesis, Properties and Applications of Oxide Nanomaterials, *John Wiley & Sons, New Jersey, USA,* 2007.
2. Xiaohong Liu, Jinqing Wang, Junyan Zhang, Shengrong Yang, *Materials Science and Engineering A* 430 (2006) 248–253.
3. Bin Cheng and Edward T. Samulski, *Chem. Commun,* (2004), 986 – 987.
4. Neenu Varghese, L.S. Panchakarla, M. Hanapi, A. Govindaraj, C.N.R. Rao, *Materials Research Bulletin* 42 (2007) 2117–2124.
5. Lanqin Tang, Bing Zhou, Yumei Tian, Hari Bala, Yan Pan, Suxia Ren, Yi Wang, Xiaotang Lv, Minggang Li, Zichen Wang, *Colloids and Surfaces A: Physicochem. Eng. Aspects* 296 (2007) 92–96.
6. A. Umar, S.H. Kim, E.-K. Suh, Y.B. Hahn, *Chemical Physics Letters* 440 (2007) 110–115.
7. Roger Mueller, Lutz M-adler, Sotiris E. Pratsinis, *Chemical Engineering Science* 58 (2003) 1969 – 1976.
8. D Michael P. Mingos , David. R. Baghurst, *Chem. Soc. Rev.,* (1991), 20, 1-47.
9. K. J. Rao, B. Vaidhyanathan, M. Ganguli, P. A. Ramakrishnan, *Chem Mater.,* (1999), 11, 882 – 895.
10. C.R. Struss, Robert W. Tainor, *Aus J. Chem.,* (1995), 48, 1665-1692.
11. ICDD File No – 05-0664 (International Center for Diffraction Data, USA).

Properties

Mater. Res. Soc. Symp. Proc. Vol. 1174 © 2009 Materials Research Society   1174-V09-07

Structural, Optical and Luminescent Properties of ZnO:Eu³⁺ Nanocrystals Prepared by Sol-Gel Method

S. Jáuregui-Rosas[1,2], O. Perales Pérez[3], W. Jia[2], O. Vásquez[2] and L. Angelats[1]

[1] Materials Physics Laboratory-Department of Physics, Universidad Nacional de Trujillo, Av. Juan Pablo II S/N, Trujillo-Perú
[2] Department of Physics, University of Puerto Rico, Mayagüez, Puerto Rico, 00681-9016, USA
[3] Department of Engineering Science and Materials, University of Puerto Rico, Mayagüez, Puerto Rico 00681-9044, USA

ABSTRACT

Highly monodisperse ZnO:Eu³⁺ nanocrystals have been synthesized by modified sol-gel method from ethanolic solutions. The effect of Eu³⁺ ions (x=0.05-0.30) concentration on the structural, optical and luminescent properties has been evaluated. No other than the ZnO-wurtzite phase was observed at all dopant levels, which was confirmed by FT-IR and Raman spectroscopy techniques. A blue shift of the exciton peak and the increase on the corresponding band gap were observed at increasing Europium contents, which would indicate an interaction between Eu³⁺ ions and the development of the ZnO host structure. The luminescence properties were also dependent on Europium contents; a systematic blue shift and enhancement of the intensity of visible luminescence peak, attributed to an increment of surface defects, was observed by a rising Europium concentration. The red luminescence band, representing the $^5D_0 \rightarrow {}^7F_2$ transition, was clearly observed in nanocrystals after annealing at 300°C for one hour. The presence of this band could be considered as an evidence of the effective energy transfer from ZnO to Eu³⁺ ions.

INTRODUCTION

ZnO, a wide direct band gap (3.37eV) with a large exciton binding energy (60meV) material, is considered as one of the most promising semiconductor material for broad application spectra [1]. In particular, nanostructured ZnO has been proposed as host structure for rare-earth (RE) ions [2-6], being the Eu-doped ZnO system widely studied in recent years as nanorods [7], thin films [8] and nanoparticles [5,9]. These structures have been prepared using diverse routes including microemulsion [7], sol-gel [5], Pechini's method [8], etc. Although several reports claimed that actual doping and energy transfer from ZnO host to Eu³⁺ ions take place [5-7], there is still controversy about the incorporation of large Eu ions into the ZnO lattice [8-12]. If possible, the synthesis of this type of doped semiconductor would open new applications due to its expected high light emission efficiency due to the Eu-Zn interactions between atomic states associated with the dopant and the host lattice material. Therefore, any effort to elucidate the formation of Eu-doped ZnO and understand the interactions between dopant and host lattice sounds justified. Accordingly, we investigated the synthesis of highly monodisperse ZnO nanocrystals containing Eu³⁺ ions by a modified sol-gel method and their structural, optical and luminescent characterization. The effect of annealing on the material functional properties is also presented.

EXPERIMENTAL
Synthesis of nanocrystals

ZnO nanocrystals, pure and doped, were prepared through a modified sol-sel method previously reported [13]. In brief, $Zn(C_2H_3O_2)_2$ and $Eu(C_2H_3O_2)_3.XH_2O$ were used as precursor materials and anhydrous ethanol as solvent. $LiOH.H_2O$ solution in ethanol was used as precipitating agent. Required amounts of acetate salts, according to $Zn_{1-x}Eu_xO$ ($x = 0.00 - 0.30$), were dissolved in 75ml of ethanol under reflux at 65°C keeping the total ion concentration at 0.1mol/L. A suitable amount of Lithium hydroxide was dissolved by sonication in 50ml of ethanol to obtain a solution at 0.14mol/L, which was added dropwise into the Zn/Eu solution and mixed under vigorous stirring for 10 minutes at room temperature. Nanocrystals were recovered by coagulation-dispersion cycles using n-hepthane and resuspended in fresh ethanol. Recovered solids were dried at 50°C for 12h and submitted for characterization.

Characterization techniques

Structural characterization was carried out in an X-Ray Diffraction (XRD) system using the Cu-Kα radiation, Fourier Transform Infrared (FTIR) spectroscopy in a MIRacleTM ATR FTS 1000 spectrometer in the transmittance mode, and Raman spectroscopy using a Renishaw micro-Raman system. The spectra were taken at room temperature using a 514.5nm excitation line from an Ar^+-ion laser. Optical properties were determined by UV-vis spectroscopy using a Beckman Coulter DU 800 spectrophotometer. Luminescence properties were determined using a Spectrofluorometer FluoroMax-2 with a continuous ozone-free Xe lamp as the excitation source.

DISCUSSION

Structural characterization

Figure 1.a shows the XRD patterns of non aged Eu^{3+}-containing ZnO nanocrystals. Except for a weak peak around 34° ($x =0.30$), which could be assigned to intermediate basic zinc acetate, no additional peaks of any other impurity were detected. All other peaks correspond to ZnO with a hexagonal wurtzite structure. These patterns are similar to those presented by Y. Liu, et al., [14] who synthesized nanocrystals of Eu^{3+}:ZnO with a nominal concentration of 2 at% following a similar synthesis approach in ethanol media. The peaks broadening are evidence of the nanocrystalline nature of the solids. The average crystallite size, which varied between 3.8nm and 3.5nm (fig. 1.b), was decreased by the presence of europium ions. This trend would indicate an inhibiting effect of Eu^{3+} ions on the growth of ZnO nanocrystals; probably, the presence of the dopants would have affected the solubility of the intermediate and hence, the dissolution - recrystallization paths conducive to the formation of the oxide. Figure 1.c shows the FT-IR spectra of the ZnO nanocrystals synthesized in presence of Eu ions. The intense IR-band centered on 524cm^{-1} can be assigned to Zn-O bound, which corroborates the formation of the host ZnO. The broad bands around 1411cm^{-1} and a 1573cm^{-1} can be assigned to asymmetric and symmetric C=O stretching vibration modes in acetate species, respectively, that could be adsorbed on them surface of ZnO.

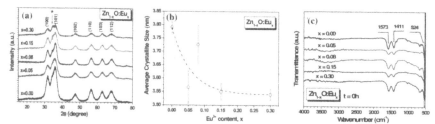

Figure 1. (a) XRD patterns of Eu:ZnO nanocrystals synthesized at various atomic fractions of Eu ions, 'x'. Only wurtzite peaks are observed. (b) Variation of average crystallite with Europium content. (c) FT-IR spectra of the same nanocrystals.

Figure 2 shows the Raman spectra of pure and Eu^{3+}-containing ZnO nanocrystals. The strongest band at $438cm^{-1}$ has been assigned to non-polar modes with E_2(high) symmetry and is associated with oxygen atom displacement. The intensity of this band increases as the Eu^{3+} ions does. The weak band at $330cm^{-1}$ was assigned to second order multiphonon processes, which originates from the zone-boundary phonon of E_2(low). The band at $581cm^{-1}$ corresponds to E_1(LO) mode and is associated with oxygen deficiency [15]. The intensity of this band increases with the content of Eu^{3+} ions, which could indicate the promoted generation of defects related to oxygen vacancies. In addition, the band assigned to the multiphonon process ($330cm^{-1}$) shifts to lower energies by rising Eu^{3+} ions contents. This trend can be due to a weak interaction between ZnO host and Eu^{3+} ions.

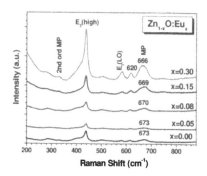

Figure 2. Raman spectra of pure and Eu^{3+}-containing ZnO nanocrystals at different 'x' values.

Optical properties

Figure 3.a presents the absorbance spectra for pure and Eu-ZnO nanocrystals. In general, the nanocrystals showed a band gap energy values above the bulk value for ZnO (3.37 eV), which can be attributed to the strong quantum confinement effect.

A well-defined exciton peak was observed for all samples that suggested a very fast ZnO formation. Furthermore, the excitonic peaks appeared at wavelength values less far below the wavelength observed for bulk ZnO (370nm). Unlike Ashtaputre's results [9], our results showed the systematic blue-shift of the exciton peaks as Eu^{3+} ions content was increased. The band gap energy (E_g) for each sample was estimated after linear fitting of the linear region of the UV-vis absorption curve, as shown in figure 3.b. Although the estimated band gape values did not exhibit a remarkable increase with rising 'x' values; however, the presence of Eu species in reacting solutions caused the nanocrystals exhibited a larger band gap energy in comparison to pure ZnO synthesized under similar conditions. As also suggested by size determination from XRD measurements (fig. 1.b), this increase in the band gap energy values in Eu:ZnO nanocrystals could be related to the inhibition on crystal growth by the presence of Eu species in reacting solutions.

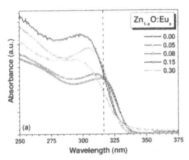 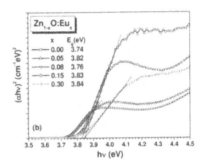

Figure 3. (a) Absorption spectra and (b) estimation of the energy gap for Eu:ZnO (x = 0.0-0.30) nanocrystals.

Photoluminescence measurements

The emission spectra corresponding to non-aged pure and Eu-containing ZnO nanocrystals powders are shown in figure 4.a. One band in the UV region and another in the green visible region were identified. The intensity of the green luminescence, which can be related to oxygen defects, is much higher than that of the exciton emission band. Moreover, the visible/UV intensities ratio increases as Eu^{3+} ions content increases, as evidenced by the data in figure 4.b. Since both emission processes compete between themselves, the trend observed in the visible/UV intensities ratio could suggest that the visible emission process may involve a step in which the photo-generated hole is trapped efficiently somewhere in the crystal; the rate of this hole trapping process must be much faster than the radiative recombination of the emission exciton. Furthermore, because of the large surface-to-volume ratio of synthesized ZnO nanocrystals, efficient and fast trapping of photo-generated holes at surface sites can be expected. The contribution of Eu^{3+} ions to the generation of defects was suggested based on the Raman spectroscopy measurements on our samples. Similar behavior was observed by Armelao et al. [16]. Figure 4 also shows that the increase on the concentration of Eu^{3+} ions was conducive to a blue shift of the green emission peak. This fact would confirm that Europium species are really inhibiting the growth of nanocrystals, in good agreement with our previous discussion based on XRD and UV-Vis analyses. These findings are also in good agreement with more recent results

[5,9,12], where the described behavior was explained in terms of the presence of europium ions on the surface of the ZnO nanocrystals.

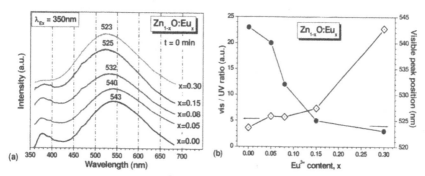

Figure 4 (a) Room-temperature emission spectra collected by using the 350 nm excitation line, and (b) variation of the visible to UV emission ratio (vis/UV ratio) and visible emission peak position with Eu concentration in Eu:ZnO nanocrystals (x = 0.0-0.30).

Annealing Effect

Eu:ZnO nanocrystals (x =0.15) were annealed at 300°C for 1 hour in air as an attempt to enhance their crystallinity and study the variation in optical properties. Figure 5 shows the corresponding XRD pattern (a), and emission spectra, (b) of as synthesized and annealed powders. As expected, the XRD spectrum of the annealed sample exhibited very sharp peaks, indicating better crystallinity than the exhibited by the as-synthesized sample. No isolated Eu-based compounds were detected. The corresponding average crystallite size increased from 3.55m (as-synthesized sample) up to 11.9nm (annealed sample).

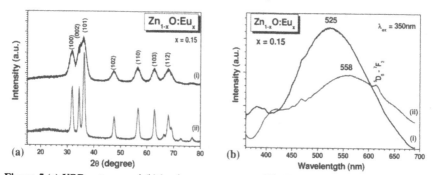

Figure 5 (a) XRD patterns and (b) luminescence spectra of Eu;ZnO, (x=0.15), as-synthesizied (i) and after annealing at 300°C for 1 hour (ii). The excitation wavelength was 350nm in the luminescence measurements.

The photoluminescence spectra for the annealed Eu:ZnO sample (figure 5.b) evidenced the red-shift of its visible luminescence band, caused by the crystal growth confirmed by XRD. The Eu^{3+} emission from $^5D_0 \rightarrow ^7F_2$ transition, clearly detected in the annealed sample, would indicate an effective energy transfer from host ZnO to dopant Eu ions. Recently, two possible models have been proposed for this behavior; the first one considers a core-shell model [9] in which nanoparticles are formed by a ZnO core and a shell of $Zn(OH)_2$ hosting the Eu^{3+} ions. The second model [5] suggests that Eu ions could have been incorporated into ZnO lattice during growth at high temperatures. Both models agreed on the formation of ZnO nanocrystals with a Eu-rich surface, which could also explain our results.

CONCLUSIONS

ZnO nanocrystals containing Eu^{3+} ions have been prepared by sol-gel method. The effect of Eu^{3+} ions on the optical absorption and photoluminescence behavior of analyzed samples clearly suggested an actual and efficient interaction between ZnO host and Eu^{3+} ions. A blue shift and simultaneous increase on the intensity of the visible luminescence peak by increasing the concentration of Eu^{3+} ions was also observed. The later would indicate an increment of surface defects due to the presence of dopant ions. The presence of the red luminescence band in the annealed sample, corresponding to the $^5D_0 \rightarrow ^7F_2$ transition, can be considered an evidence of an effective energy transfer from ZnO to Eu^{3+} ions.

ACKNOWLEDGMENTS
This material is based upon work supported by the National Science Foundation under Grant No. 0351449. OJPP also acknowledges the support from the DoE-Grant No. FG02-08ER46526

REFERENCES
- [1] Ü. Özgür, et al., J. Appl. Phys. **98** (2005) 041301
- [2] S.M. Liu, et al., Phys. Lett. A **271** (2000) 128-133
- [3] H.L. Han, et al., Opt. Mater. **31** (2008) 338-341
- [4] Y. Bai, et al., Opt. Commun. **281** (2008) 5448-5452
- [5] Y. Liu, et al., J. Phys. Chem. C **112** (2008) 686-694
- [6] Y. Zhang, et al., J. Phys. D: Appl. Phys. **42** (2009) 085106
- [7] A. Ishizumi and Y. Kanemitsu, Appl. Phys. Lett. **86** (2005) 253106
- [8] S.A.M. Lima, et al., Appl. Phys. Lett. **90** (2007) 023503
- [9] S.S. Ashtaputre, et al., J. Phys. D: Appl. Phys. **41** (2008) 015301
- [10] W. Jia, et al., Opt. Mater. **23** (2003) 27-32
- [11] A.A. Bol, et al., Chem. Mater. **14** (2002) 1121-1126
- [12] L. Robindro Singh, et al., J. Lumin. **128** (2008) 1544-1550
- [13] O. Perales-Pérez, et al., Nanotechnology **18** (2007) 315606
- [14] Y. Liu, et al., Opt. Lett. **32** (2007) 566-568
- [15] S. Gao, et al., Appl. Phys. Lett. **89** (2006) 123125
- [16] L. Armelao, et al., J. Phys. Chem. C **111** (2007) 10194-10200

Mater. Res. Soc. Symp. Proc. Vol. 1174 © 2009 Materials Research Society   1174-V05-01

Effect of Morphology of ZnO Nanowire Arrays on Interfacial Shear Strength in Carbon Fiber Composites

Ulises Galan[1], Gregory J. Ehlert[1], Yirong Lin[1] and Henry A. Sodano[1]
[1]Mechanical and Aerospace Engineering, Arizona State University, Mail Stop 6106
Tempe, AZ 85287-6106, U.S.A.

ABSTRACT

ZnO nanowire arrays have been grown on carbon fiber to enhance the interface strength of a polymer matrix composite without degrading the base fiber and in-plane strength of the composite. The morphology of the nanowire array is controlled during growth to create nanowires with different aspect ratios to elucidate the structure-property relations of the interphase. Nanowires are shown to double the composite interfacial shear strength at an intermediate nanowire length, indicating that an optimal point exists and the interface can be engineered to maximize the interfacial enhancement. Furthermore, the observed effect of the morphology on interface strength indicates that the bond between the ZnO nanowire array and the carbon fiber is quite strong, more than twice as strong as the interaction between the matrix and control fiber.

INTRODUCTION

The continued development of higher performance structures and systems relies heavily upon the employment of advanced composite materials. Composite materials have become a common choice because they offer high specific strength, and toughness as well as multiple design parameters to tailor the material to the specific structural demands. Carbon fiber epoxy matrix composites have attracted significant academic and commercial attention because they have some of the highest specific properties of commonly available fibers and matrices[1].

It has been shown that the bulk properties of a carbon fiber-epoxy composite system depend heavily on the interface between the fiber and matrix[2]. As such, improving the interface has garnered significant attention. One approach is to add a new interphase that interacts strongly with both the fiber and matrix, effectively mechanically binding the two. Many different interphase designs have been attempted that bond with the fiber and mechanically interact with the matrix; such as SiC whiskers[3], carbon nanotube arrays[4-7] and carbon nanofiber arrays[8]. While these have improved interfacial shear strength, they suffer from poor fiber – interphase bonding and/or degradation of the base fiber due to the high temperatures and catalysts required for growth[6]. Furthermore, these processes have not investigated the effect of changing the geometry of the surface enhancement in order to optimize or maximize the effect of the interphase[10]

ZnO nanowire arrays have been used in many applications such as power harvesting[11], nano-lasers[12], solar cells[13] and chemical sensing[14]. ZnO nanowire arrays have also been shown to be compatible with many substrates including polymers[15]; however this investigation is one of the early works where the substrate is carbon fiber, a structural composite reinforcement. ZnO nanowire arrays offer substantial advantages compared to other previously investigated whiskerization techniques because the low temperature hydrothermal process does not damage the in-plane properties or the thermal stability of the composite[9]. Lin et al also demonstrated that the technology can be extended to lamina scale and demonstrated a 40% increases in the interfacial shear strength and shear modulus. Furthermore, the hydrothermal growth procedure is compatible with large scale reel to reel manufacturing, making this a promising commercial technology for enhancement of polymer matrix composites.

Lin et al[9] observed very large increases in interfacial shear strength due to increased surface area and mechanical interlocking of the nanowires with the impregnated matrix; however there were only two cases tested, with and without nanowire arrays. It is suspected that shear strength should be a function of nanowire length; short nanowires would be similar to surface roughness or even bare fibers, whereas long nanowires should provide maximum surface area and mechanical interlocking to enhance the interfacial shear strength. While longer nanowires are expected to enhance the interface, this study will focus on the effect of nanowire length, in particular if an optimum nanowire length exists. The use of ZnO nanowires allows the arrays to be precisely tuned by adjusting the growth parameters to create many lengths and in order to investigate the effects of nanowire length on interfacial strength. Furthermore, if interfacial shear strength changes are detected by changing the nanowire morphology, failure is likely occurring at the nanowire – matrix interface and not the nanowire – fiber interface. This information will highlight the weakest link in the system and enable focused study for further improvement of the interfacial shear properties.

EXPERIMENT

ZnO nanoparticles are grown in an ethanol suspension by first creating a 0.0125M solution of zinc acetate dihydrate (Reagent Grade, Sigma) and 0.02M solution of sodium hydroxide (ACS Grade, EMD Chemicals). 80 mL aliquots of the zinc acetate – ethanol solution and the sodium hydroxide – ethanol solution are then diluted with 640 mL and 200 mL of pure ethanol, respectively. The two diluted solutions are then heated to 55°C, mixed under vigorous stirring and maintained at the growth temperature of 55°C for 30 minutes. The seed suspension is then quenched in ice water down to room temperature and found to stay clear for up to 2 weeks, consistent with previous studies[16].

Unsized IM7 carbon fibers (Hexcel, Stamford CT) are first cleaned in ethanol and acetone. The fibers are then dip coated in the ZnO nanoparticle suspension and annealed for ten minutes at 150°C three times. ZnO nanowires are grown from the deposited seed layer on carbon fibers by putting the fibers in a preheated aqueous solution of 0.025M zinc nitrate hexahydrate (99%, Alfa Aesar), 0.025M hexamethylenetetramine (99%, Alfa Aesar) and 0.005M branched polyethylenimine (PEI) (Sigma, M_w ~25,000) at 90°C in water bath. The solution is replaced every 2.5 hours and the fibers are washed with deionized water each time the solution is changed. Nanowire morphology was analyzed in a Hitachi S4700 FESEM at 5.0 keV.

Single fibers are then placed into silicone rubber (3120 RTV, Dow Corning Corp, Midland MI) dog bone molds and filled with Epon 862 (Hexion Inc, Houston TX) and Epikure 9553, mixed 100:16.9 parts by weight. The epoxy was gelled for 90 minutes at 45°C on a hotplate, followed by 1 hour in an oven at 100°C and 1 hour at 160°C. The samples were then polished to 2400 grit for testing in a custom built micro-tensile frame under polarized transmitted light; similar to Feih[17]. Strain was incrementally applied and the fiber scanned for cracks. The fibers were continually strained until the number of cracks saturated and did not change with applied strain. Strain rate was approximately 0.6% / min during straining and approximately 3 minutes was given to count the number of cracks in the sample.

DISCUSSION

Growth of nanowire arrays on carbon fiber

The morphology of the ZnO nanowire arrays were controlled through variation of the growth time as well as the addition of PEI to the growth solution. The resulting nanowire arrays for each of the growth conditions tested here are shown in Figure 2, as grown under various conditions and times on carbon fiber tows. The SEM micrographs show only a single fiber; however the images are representative of the coating on the surface of all the fibers in the tow. It should also be noted that the fiber shown in Figure 2 are the same as those tested. We observed that the nanowire arrays often would not grow on the internal fibers of a woven fabric, because the ions for crystal growth would get consumed before diffusing into the center of the fabric. We also observed that as growth time increased, a slightly longer nanowire would penetrate more into open solution and grow faster than its shorter neighbors. This is observed by an increasing standard deviation of nanowire length over the fiber which is shown in Figure 1.

In addition to measured lengths of the nanowires, the images in Figure 2 also show the growth of larger crystalline ZnO that are not adhered to the fiber but have precipitated out of solution and deposited on the fibers. It was observed that excessive 'flowering' or crystallites significantly degraded the fiber interface strength and therefore great care was taken to both remove the crystals through washing and ensure that only fibers with minimal crystals were tested. Typically, several washes in deionized water both during and after growth of the nanowires were required to sufficiently remove the crystals. Sonication also removed the crystals very effectively; however sonication

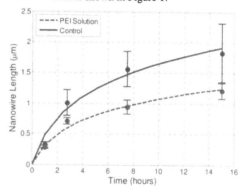

Figure 1. Growth results of ZnO nanowire arrays on carbon fiber. A logarithmic growth model is fit and shows good agreement with the measured data. R^2 values are at least 0.95 for both control and PEI.

also damaged the underlying nanowire arrays and thus reduced the interface strength of the composite.

PEI was employed by Law et al[13] to reduce the radial growth rate of the nanowires and allow for extremely long, high aspect ratio nanowire arrays. PEI was not observed in this testing to significantly inhibit radial growth, but rather reduced the longitudinal growth rate of nanowires, as seen in Figure 1. It was observed to maintain better alignment of the nanowire arrays (Figure 2G and 2H) and reduce the presence of longitudinal or randomly oriented nanowires. This contradicts previous findings[13]; however it does validate that polymer surface additives can be used to further control the growth of ZnO nanowire arrays. Other polymers, temperatures or growth times could be used to control the diameter and aspect ratio of the nanowires.

For the experiments performed here, empirically derived control formulas were required so that various nanowire lengths could be created for interface testing. The eight images in Figure 2

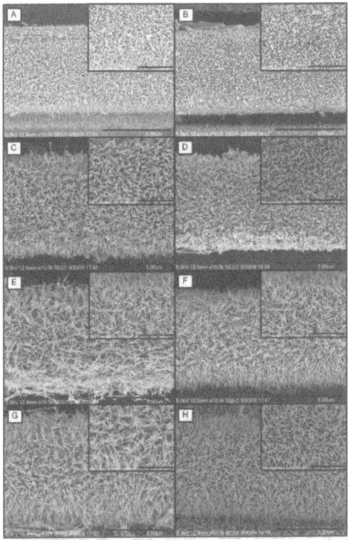

Figure 2. Growth of ZnO nanowire arrays on IM-7 carbon fiber. Growth times for A and B were 1 hour; C and D, 2.75 hours; E and F, 7.5 hours; G and H, 15 hours. A, C, E, and G were produced in the control solutions. B, D, F, and H were produced in a similar solution with the addition of polyethyleneimine to slow radial growth. All primary scale bars are 5 μm and all inset scale bars are 2 μm.

were then measured in 11 random spots for nanowire length and the corresponding averages with confidence intervals were plotted. The measured data is presented as means with error bars in Figure 1. It is observed that the standard deviation does increase with time and that the length also increases for each successive observation. Each of these nanowire lengths is then tested for the corresponding interface strength.

Interface testing

Single fiber interface testing was performed to evaluate the effect of morphology on the interfacial properties of the composite. The goal of single fiber fragmentation is to measure the average critical length of the fragments which can be used to determine the interface strength. If the base fiber strength is degraded, the number of cracks will also increase; however Lin et al showed that the carbon fibers maintain tensile strength after the growth procedure. In addition, testing performed in this investigation found that too high of an incremental strain rate would not saturate the fiber to the fullest extent; thus a very slow strain rate was applied to the specimens.

Adding a new interphase to the composite replaces one interface with two new ones, and consequently both must be stronger than the original to yield increases in net strength. The first new interface is between the interphase and the structural fiber; in this study between carbon and ZnO. It has been observed that ZnO interacts strongly with carboxylic acid functional groups[18-19] and it is known that carbon fiber surface is often coated with many such functional groups[20]. If failure is occurring at the nanowire – fiber interface, then morphology will have no effect on the interfacial shear strength. If an increase is detected by changing the nanowire length, then failure is probably occurring at the interface of the nanowires and matrix, which yields insight into which interface needs to be studied for maximum interfacial shear strength.

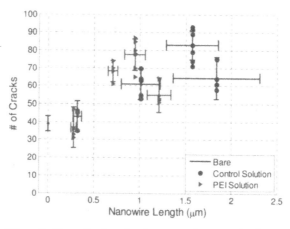

Figure 3 shows the number of cracks in each sample as related to the nanowire length. The number of cracks is shown in Figure 3 to increase from 39

Figure 3. Strengthening of polymer matrix composite due to functional gradient and load transfer enhancement. A local maximum exists relative to nanowire length. The definite effect of nanowire length implies that the interphase bond with the base fiber is at least as strong as the nanowire – matrix interaction.

to 83, an increase of 112%. It is clear that a maximum does exist for both solutions at the second longest growth length. As hypothesized, the addition of longer nanowires does improve the interface strength of the composite; however the nanowires should not be excessively long. The addition of longer nanowires is suspected to generate a functional gradient of properties and reducing the stress concentration between the fiber and matrix. Longer nanowires also appear to reduce the uniformity of the nanowire arrays, which may cause stress concentrations and reduce load transfer. The effect of the morphology validates the hypothesis that bonding between the nanowire arrays and the fiber is at least as strong as the matrix – nanowire interaction; otherwise interfacial strength would be independent of the nanowire – matrix interface. In light of the effects of length on the interfacial strength of the composite, it is suspected that diameter and aspect ratio also play a role. Future work includes quantifying the growth parameters to control nanowire diameter and aspect ratio and then determining the effects of all 3 morphology parameters on the interfacial strength.

CONCLUSIONS

The interfacial strength is observed to increase by 112%, which could usher in a new era of even tougher, stronger composite materials. This investigation of interphase morphology effects not only shows that the interphase can improve interfacial strength but it also shows that an optimal length of nanowire reinforcement exists. Since the morphology changes affect the interfacial strength, the interface between the nanowires and matrix is believed to be the weakest link. This indicates that the nanowire – matrix interactions must be studied to further enhance interphase strengthened composites.

ACKNOWLEDGMENTS

The authors would like to gratefully acknowledge support from the Army Research Office (Award # W911NF0810382) and the National Science Foundation (Grant # CMMI- 0846539)

REFERENCES
[1] I. Daniel and O. Ishai, Engineering Mechanics of Composite Materials, Oxford Univ. Press, Oxford, 2006.
[2] M. W. Hyer, Stress Analysis of Fiber Reinforced Composites Materials, McGraw- Hill, New York, NY, 2001.
[3] W. Kowbel, C. Bruce, J. Withers, and P. Ransone, Comp. A, 28, 993-1000 (1997).
[4] R. Sager, P. Klein, D. Lagoudas, Q. Zhang, J. Liu, L. Dai, and J. Baur, Comp. Sci. Tech, In press (2009).
[5] L. Qu, Y Zhao, and L. Dai, Small, 2, 8, 1052-1059 (2006)
[6] E. Thostenson, W. Li, D. Wang, Z. Ren, and T Chou, J. App. Phys., 91, 6034 (2002).
[7] E. Thostenson, C. Li, and T. Chou, Comp. Sci. Tech., 65, 3, 491 (2005).
[8] B. Boskovic et al, Carbon, 43, 13, 2643 (2005).
[9] Y. Lin, G. Ehlert, and H. Sodano, Adv. Func. Mater., In Press (2009).
[10] A. Esawi and M. Farag, Mater. Design, 28, 9, 2394-2401 (2007).
[11] X. Wang, J. Song, J. Liu, Z. Wang, Science, 316, 5821, 102-105 (2007).
[12] Z. Tang et al, Appl. Phys. Lett., 72, 3270 (1998).
[13] M. Law, L. Greene, J. Johnson, R. Saykally, and P. Yang, Nat. Mater., 4, 455-459 (2005).
[14] Q. Wan et al, Appl. Phys. Lett., 84, 3654 (2004).
[15] L. Greene, M. Law, D. Tan, M. Montano, J. Goldberger, G. Somorjai, and P. Yang, Nanoletters, 5, 7, 1231 (2005).
[16] Z. Hu, G. Oskam, and P. Searson, J. Col. and Int. Sci, 263, 454-460 (2003).
[17] S. Feih, K. Wonsyld, D. Minzari, P. Westermann, and H. Liholt, Riso-Report, R-1483(EN) (2004).
[18] G. Wegner, P. Baum, M. Muller, J. Norwig and K. Landfester, Macromol. Symp., 175, 349 (2001)
[19] Ravindran and C. Ozkan, Nanotechnology, 2005, 16, 1130-1136.
[20] E. Palma and L. Ibarra, Makro. Chemie, 220, 111-122 (1994).

Mater. Res. Soc. Symp. Proc. Vol. 1174 © 2009 Materials Research Society   1174-V06-25

Evolution of Textural, Structural and Morphological Properties of Ag/TiO2 Nanocomposites Tailored by Temperature

Marcelo M. Viana[1], Nelcy D. S. Mohallem[1,2], Douglas R. Miquita[2] and Karla Balzuweit[2]
[1]Laboratório de Materiais Nanoestruturados, UFMG, Belo Horizonte, Brazil
[2]Centro de Microscopia, UFMG, Belo Horizonte, Brazil
nelcy@ufmg.br

ABSTRACT

TiO2 is a promising material for use in environmental purification due to its strong oxidizing power under UV illumination, non-toxicity and long-term photostability. Nanocomposites formed by silver nanoparticles dispersed in titania matrix have their application quality improved since silver particles can act in the electronic structure of the titania. In this work, Ag/TiO2 nanocomposites were prepared by sol-gel process. The structural evolution of the nanoparticles and the dependence of particle size with the calcination temperature were investigated.

INTRODUCTION

Nanometric TiO2 systems are used in photocatalysis [1-4], material sterilization, superhydrophilic surfaces [5], gas sensors [6], superconductivity [7], among others applications. On the other hand, the silver has been used as a bactericidal material since the antiquity and in the current days nanoparticulate silver is used in commercial materials, adding value to the final product [8]. Ag/TiO2 nanocomposite has its potentiality of application improved when compared with the Ag and TiO2 in their simple forms, because silver particles act as electron traps aiding electron–hole separation, inducing the photocatalysis in the visible region [9].

Ag/TiO2 nanostructured have been prepared by various methods of synthesis as hydrothermal process, used in the production of nanotubes and nanosheets [10], spray pyrolysis producing spherical nanoparticles [11], beyond methods that generate core-shell nanostructure [12], nanowire [13] and nanofibers [14]. Besides, considerable efforts have been made in studies involving thin films with bactericide [15] anticorrosive [16] and photochromic [17] properties.

In this work, titanium isopropoxide and silver nitrate solution was used as precursor of Ag/TiO2 nanocomposites. After irradiation and gelation of the precursor solution at room temperature, this material was dried and calcined at various temperatures up to 1100 °C. The evolution of textural, structural and morphological properties of the nanocomposites with the calcination temperature was studied.

EXPERIMENT

A solution with titanium isopropoxide (IV) (Ti(OCH(CH3)2)4) 97% (Aldrich) diluted in isopropyl alcohol 99.5% (Merck) mixed with acetic acid was magnetically stirred for 1 hour. Silver nitrate diluted in acetonitrile was added to the starting solution, with Ag:Ti ratio of 1:6. The final solution was conditioned in a chamber and submitted to stirring and ultraviolet irradiation UV-C (254 nm) for 12 hours, using two fluorescent light-

bulbs of mercury (Girardi RSE20B) of 15 W each one. The solution gelified after 10 days, being then submitted to thermal treatment between 100 and 1100 °C, during 2 h.
The samples were characterized by thermogravimetry-differential thermal analysis (TG/DTA) in a TA Instrument (SDT 2960) in air atmosphere until 1200 °C at 10 °C.min^{-1}. The samples were investigated by X-ray diffraction (RIGAKU Geigerflex model- 3034), scanning electron microscopy (SEM, type Quanta 200 FEG - FEI) with an accelerating tension of 30 kV and high-resolution transmission electron microscopy (HRTEM) using a FEI TECNAI G2-20 microscope at acceleration tension of 200 kV. Textural characteristics of the materials were investigated by gas adsorption in an Autosorb – Quantachrome NOVA 1200.

DISCUSSION

The simultaneous TG/DTA curves of the as-prepared Ag/TiO$_2$ nanocomposite are shown in Figure 1. The DTA curve exhibits five events, two endothermic and three exothermic. The endothermic peak at 68 °C corresponds to 15% weight loss of nondissociatively adsorbed water molecules as well as water held on the surface by hydrogen bonding. In the 140 – 540 °C range, it is observed three exothermic events related to 14% weight loss, corresponding to dehydroxylation process and combustion of carbonaceous residues. An endothermic event related to anatase-rutile transition can be observed above 790 °C. The large endothermic event above 800°C is characteristic of the densification process corroborated by SEM results. Both events are overlapped.

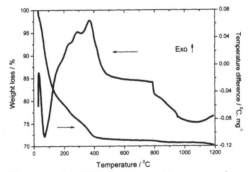

Figure 1 - TG/DTA curves of Ag/TiO$_2$ nanocomposite as-prepared.

Figure 2 shows the characteristic X-ray diffractograms of Ag/TiO$_2$ prepared at different temperatures. The sample treated at 100 °C exhibit a broad peak related to nanocrystaline anatase phase. It is observed an increase in the intensity and narrowing of the peaks with increasing temperature, indicating an increase in the crystallite size. XRD peaks related to (1 1 1) reflection of the silver can be observed in temperatures above 400 °C. A partial phase transition of anatase to rutile occurs at 700 °C, which is complete at 900 °C. At 1100 °C only rutile and silver phase is observed.

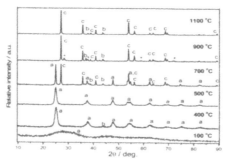

Figure 2. Diffractogram of Ag/TiO$_2$ calcined between 100 and 1100 °C (a=TiO$_2$ anatase, b=silver and c= TiO$_2$ rutile)

SEM micrographs of irradiated Ag/TiO$_2$ nanocomposites calcined between 100 and 1100 °C are shown in Figure 3. The samples treated at 100 °C are formed by fragments containing nanoparticles in the surface with average size of 80 nm. The samples calcined at 400 °C are formed by encapsulate nanoparticles with average size of 40 nm. The samples calcined at 500 °C have heterogeneous morphologies with different forms beyond nanoparticles with average size around 50 nm. The sample calcined at 700 °C exhibits nanoparticles of size around 15 nm imbibed in a matrix. In this temperature the sample has two titania phase, according XRD diffractogram. At 900 °C we can observe an

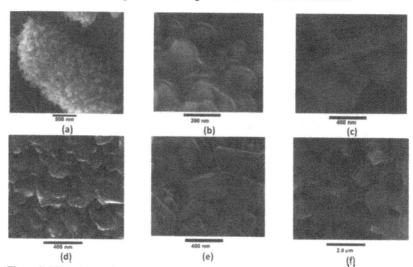

Figure 3- SEM micrographs of irradiated Ag/TiO$_2$ nanocomposite calcined at (a) 100 °C, (b) 400 °C, (c) 500 °C, (d) 700 °C, (e) 900 °C e (f) 1100 °C.

oriented growth in a prismatic structure (rutile phase) with average size particles of 15 nm dispersed in the surface. A densification process occurs at 1100 °C without the visual presence of nanoparticles in the surface.

High resolution TEM images obtained from Ag/TiO$_2$ samples calcined at 400 °C (Fig. 4) were used to determine the interplanar distance of the crystalline phases. The crystallographic parameters were obtained by Crystallographica Search-Match using XRD patterns as listed: TiO$_2$ (anatase) PDF4-477, TiO$_2$ (rutile) PDF21-1276, TiO$_2$ (brookite) PDF2-514, Ag PDF2-1098, Ag$_2$O PDF3-796.

The Fourier transform (FFT) of HRTEM images (Fig. 4a and 4c) are shown in Figures 1b and 1d, respectively. The polycrystalline nature of the nanoparticles is better illustrated in Figure 4b that shows diffraction spots related to the atomic spacing d =3.51 Å arranged in ring form. This d spacing coincides with (1 0 1) anatase plane. In this figure we can also observe two spots coinciding with the metallic silver plane (1 1 1), that have d = 2.35 Å. The Figures 4c and 4d show a interplanar distance of 3.25 Å related to (1 1 0) crystalline rutile plane. Fig. 4f shows the brookite crystalline phase identified by the interplanar distance d = 3.46 Å related to (1 1 1) plane. The average particle size of anatase, rutile, brookite and silver nanoparticles present in this sample measured by TEM is (8 ± 2), (11 ± 2), (10 ± 2) and (3 ± 1) nm, respectively.

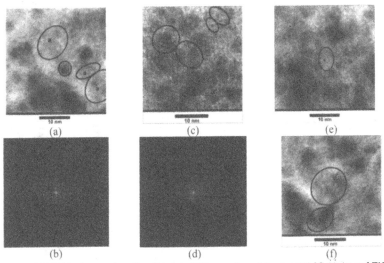

Figure 4- HRTEM micrographs of irradiated Ag/TiO$_2$ samples calcined at 400 °C (a) Ag and TiO$_2$ nanoparticles (anatase), (b) FFT of (a), (c) TiO$_2$ nanoparticles (rutile), (d) FFT of (b), (e) TiO$_2$ nanoparticle (anatase) and (f) TiO$_2$ nanoparticle (anatase and brookite).

The HRTEM results are in discordance with XRD ones, which show only the presence of metallic silver and anatase phases. These facts suggest that the electron beam with tension of 200 kV induced crystalline phase transformation of TiO$_2$ anatase phase in brookite and rutile. These transformations were visually observed during the experiments.

Some textural parameters of Ag/TiO$_2$ nanocomposites in function of the calcination temperature are shown in Table 1 and their respective adsorption-desorption isotherms are showed in Fig. 5. The sample calcined at 100 °C has higher specific surface area (389 m^2.g^{-1}) than other samples, and exhibits type I isotherm, characteristic of microporous materials (Langmuir theory), according to the BDDT classification. The C parameter or BET constant, which shows the degree of interaction between adsorbent and adsorbate had acceptable values, according literature [18].

Table 1- Textural characteristics of samples calcined at various temperatures. S$_{BET}$: specific surface area (mesopores), S$_{Langmuir}$: specific surface area (micropores), P$_{BJH}$ mesoporosity.

Temperature °C	S$_{Langmuir}$ m^{2}g^{-1}	S$_{BET}$ m^{2}g^{-1}	S$_{total}$ m^{2}g^{-1}	Constant C (BET)	P$_{BJH}$ %
100	389	-	389	-	-
400	60	84	144	134	47
500	-	95	95	125	7
700	-	7	7	174	-
900	-	3	3	181	-
1100	-	1	1	300	-

Error: 7 %

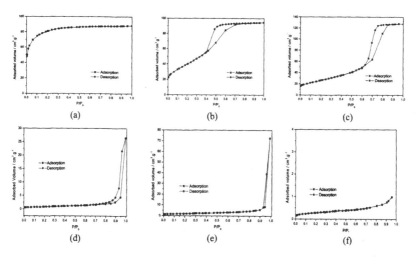

(a)   (b)   (c)

(d)   (e)   (f)

Figure 5- Adsorption-desorption isotherms of Ag/TiO$_2$ samples calcined at (a) 100 °C, (b) 400 °C, (c) 500 °C, (d) 700 °C, (e) 900 °C and (f) 1100 °C.

The texture of the nanocomposites changed with the increase in the temperature, due to evaporation of solvents and the densification process. These processes induce an increase in the pore size but a decrease in the pore quantity. The samples calcined at 400 and 500 °C presented type IV isotherms, characteristic of mesoporous material according to the BDDT classification [18] and specific surface area of 144 m^2g^{-1} (60 m^2g^{-1} related to micropores and 84 m^2g^{-1} to mesopores) and 95 m^2g^{-1}, respectively. The sample calcined at 500 °C adsorbed 127 $cm^3.g^{-1}$ of gas, the higher value than other samples, due to its minor porosity. The specific surface area and porosity decreased considerably above 700 °C due to the phase change and densification process, and the samples presented type V isotherms, characteristics of macroporous or non-porous materials.

CONCLUSIONS

The samples calcined at 100 °C are formed by nanocystalline anatase and organic solvents. Between 400 and 500 °C the samples are constituted by nanocrystalline anatase phase, are mesoporous and consequently are adequate to photocatalytic application. The nanocomposite calcined at 700 °C present an anatase-rutile phase mixture but with small specific surface area. At temperatures higher than 900 °C only the rutile phase is observed, with an increasing in the average particle size and a decreasing in the specific surface area. The sample calcined at 400 °C presented a mixture of anatase, rutile and brookite crystalline phases when observed by HRTEM. These results show the large sensitivity of the samples to the electron beam.

ACKNOWLEDGMENTS

This work was supported by FAPEMIG and CNPq (Brazilian funding agencies). The authors thank the Microscopy Center of UFMG by SEM and HRTEM images.

REFERENCES

1. M.N. Michael, K.H.Leung, Y.C.Dennis, and K.S. Leung, *Ren. Sustain. Energy Rev.* **11**, 401–425 (2007).
2. M. Nag, P. Basak, and S. V. Manorama, *Mat. Res. Bull.* **42**, 1691–1704 (2007).
3. T. L. Hsiunga, H. P. Wanga, and H. Ping, *J. Phys. Chem. Sol.* **69**, 383–385 (2008).
4. G. Sivalingam, and G. Madras, *Appl. Catal. A* **269**, 81 – 90 (2004).
5. N. Masahashi, S. Semboshi, N. Ohtsu, and M. Oku, *Thin Solid Films* **516**, 7488–7496 (2008).
6. L. Castaneda, and M. Terrones, *Physica B* **390**, 143–146 (2007).
7. V. Dallacasa, F. Dallacasa, *Physica C* **468**, 781–784 (2008).
8. S. M. Magaña, P. Quintana, D. H. Aguilar, J. A. Toledo, C. Angeles-Chavez, M. A. Cortés, L. Léon, Y. Freile-Pelegrín, T. López, R. M. T. Sánchez, *J. Mol. Catal. A* **281**, 192–199 (2008).
9. K. S. Michael, R. George, F. Patrick, and C. P. Suresh, *J. Photochem. Photobiol. A* **189**, 258–263 (2007).
10. Y. Lai, Y. Chen, H. Zhuang, and C. Lin, Mater. Lett. **62**, 3688–3690 (2008).
11. K. Kaneko, W. Moon, K. Inoke, Z. Horita, S. Ohara, T. Adschiri, H. Abed, and M. Naito, *Mater. Sci. Eng. A* **403**, 32–36 (2005).
12. L. Zhang, D. Xia, and Q. Shen, *J Nanopart Res* **8**, 23–28 (2006).
13. J. Du, J. Zhang, Z. Liu, B. Han, T. Jiang, and Y. Huang, *Langmuir* **22**, 1307-1312 (2006).
14. A. Borras, A. Barranco, and A. R. Gonzalez, *Langmuir* **24**, 8021-8026 (2008).
15. M. P. Reddy, A.Venugopal, and M. Subrahmanyam, *Water Res.* **41**, 379 – 386 (2007).
16. F.X. Perrin, V. Nguyen, and J.L. Vernet, *Polymer* **43**, 6159–6167 (2002).
17. K. L. Kelly, and K. Yamashita, *J. Phys.Chem. B* **110**, 7743-7749 (2006).
18. S. LOWELL and J. E. SHIELDS, *Powder Surface Area and Porosity*, 3rd ed. (Champman & Hall, Australia, 1991).

Mater. Res. Soc. Symp. Proc. Vol. 1174 © 2009 Materials Research Society   1174-V06-07

Rectifying Polarity Switch of Pt/TiO$_{2-x}$/Pt

Ni Zhong[1,2], Hisashi Shima[1,2], and Hiro Akinaga[1,2]

[1] Nanotechnology Research Institute (NRI), National Institute of Advanced Industrial Science and Technology (AIST)
[2] CREST, Japan Science and Technology

ABSTRACT

Current-voltage *(I-V)* characteristic of Pt/TiO$_{2-x}$/Pt has been investigated. The Pt/TiO$_{2-x}$/Pt devices in the initial state exhibit a rectifying *I-V* behavior. By applying a pulse voltage, the rectifying polarity could be switched to an opposite direction. The mechanism of the rectifying polarity switch is proposed as the local drift of defects, such as oxygen vacancies (V_O), due to applying pulse voltage. It is found that the required pulse voltage height for the polarity switch (V_{switch}) exhibits much dependence on the operation temperature and width of applied pulse voltage. With an increase of the pulse voltage width or the measurement temperature (T), V_{switch} exhibits a decrease with increase of T. These results suggest that the rectifying polarity switch in the Pt/TiO$_{2-x}$/Pt is attributed to a thermal and dynamic dependence process, which agree well with the localized migration of V_O induced by applied pulse voltage.

INTRODUCTION

TiO$_{2-x}$ attracted considerable interest in the application of heterogeneous photocatalysts, solar cells, gas sensor, biocompatible materials and nanoscale electronic device. Most recently, current conduction and carrier transport of the TiO$_{2-x}$/metal interface are considered to be applicable in the semiconductor technology.[1,2] In our group, a novel switchable property has been explored on a simple metal/oxide/metal (MOM) structure by carefully controlling the fabricating process.[3,4] This switch device could be considered as a simple diode. And the polarity of this diode is switchable by applying pulse voltage even after it is prepared. TiO$_{2-x}$ could be considered as a doped n-type semiconductor because of oxygen vacancy. A direct observation of the field-induced motion of color centers in rutile crystals has been reported.[5,6] By applying a field, blue color drifts towards negative terminal, and positive terminal becomes transparent.[6] It is attributed to the motion of V_O due to the applied field. Moreover, It has been reported that formation of a continuous channel of V_O was revealed by X-ray absorption near-edge spectroscopy in SrTiO$_3$.[7] It could explain a sudden current jump after a forming process found in SrTiO$_3$ crystal with Pt electrodes,[8] which is also found in rutile crystal.[9] Therefore, by applying pulse voltage, motion of V_O is expected as a thermal and dynamic dependence process in oxide semiconductor such as TiO$_{2-x}$ and SrTiO$_3$.

The mechanism of the switch characteristic in our prepared Pt/TiO$_{2-x}$/Pt device has been attributed to the shunting and recovery of the metal/oxide interfacial electronic barrier due to the localized drift of V_O. Most recently, the Schottky barrier at Pt/TiO$_{2-x}$ interface has also been probed by electric beam induced current (EBIC), which is support for the proposed mechanism. And it will be reported in the future. As known, the motion of V_O is thermally activated, and applied electric force, especially pulse voltage width (w) may also play a role in the switch behavior. Therefore, in the present work, the dependence of applied electric force and operating temperature on the switch behavior was studied to obtain further information of this novel switch characteristic.

EXPERIMENT

TiO_{2-x} thin films with a thickness of 50 nm were grown on the substrate with a structure of $Si/SiO_2/Ta(5nm)/Pt(100nm)$ by radiofrequency (rf) sputtering. Pt has been selected as electrode due to its high work function. For the reactive sputtered TiO_{2-x} layer, the partial pressure of Ar and O_2 during sputtering is 0.45 and 0.05 Pa, respectively. The rf power was 200W. Rectangular shaped top electrodes Pt with the thickness of 100nm were micro fabricated by the conventional photolithography process. The dc I-V characteristic is acquired with a Keithley 4200 semiconductor parameter analyzer. Applying pulse voltage was driven at top electrode (TE), and bottom electrode (BE) was grounded in all the measurement.

RESULTS AND DISCUSSION

Figure 1(a) shows the initial and after applying pulse current-voltage (I-V) curves. In the initial state, the forward current was observed in the positive bias. After applying a negative pulse voltage, the forward current was observed in the negative bias, for example, the I-V curves after -7.0V/900ms, -8.0V/900ms and -10V/900ms. It demonstrates that the diode polarity was switched. Figure 1(b) show the dependence of read current at ±2.0 V on applying pulse voltage height. With increase of pulse voltage height, forward current decreases, while reverse current increases. When reverse current becomes larger than forward current, the polarity of the $Pt/TiO_{2-x}/Pt$ diodes switches. The pulse voltage height required for polarity switch is called V_{switch}, hereafter, as sign in the Fig. 1(b).

(a)   (b)

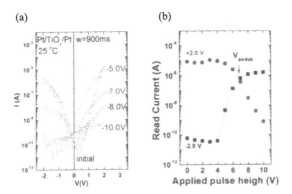

Fig. 1 I-V characteristics of $Pt/TiO_{2-x}/Pt$ in the initial state and after applying pulse voltage, pulse width w: 900ms (a).Current at ±2.0V as a function of the applied pulse height (b).

In the previous work, the mechanism of the polarity switch behavior is proposed as the electrical control of the defects such as V_O at the Pt/TiO_{2-x} interface.[3,4] As known, motion of defect such as V_O is a thermal activated process. Therefore, effect of the operation temperature, and width of pulse voltage on the switch characteristic, especially on the V_{switch} was studied to probe the assumed mechanism.

As shown in Fig.2 (a) and 2(b), with increase of the temperature, a decrease of V_{switch} is found. Increase of the applied pulse width also results in the decrease of the V_{switch}. In a word, high temperature and long width of pulse voltage is beneficial of the switch of the Pt/TiO$_{2-x}$/Pt. To explain the influence of temperature on V_{switch}, we could like to mention the concept of the diffusion coefficient of defect, which can be expressed as:

$$D(T) = D_0 exp \left(\frac{-E_a}{kT}\right) \tag{1}$$

where D_0 is the pre-exponential factor, T is the absolute temperature, E_a is the activation energy for diffusion, and k is Boltzmann's constant.

On the basis of Equ.(1), it is clear that high temperature results in high value of diffusion coefficient, while low temperature contributes to low value of diffusion coefficient. It agrees well to the present results high operation temperature favor the switch behavior of the Pt/TiO$_{2-x}$/Pt devices.

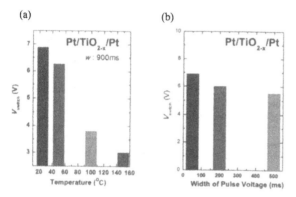

Fig. 2 The dependence of V_{switch} on the operation temperature (a); The dependence of the V_{switch} on the width of pulse voltage.

The mechanism of the polarity switch is considered as following:

(1) Initial state: a Schottky barrier forms at TiO$_{2-x}$/Pt TE interface, while Pt BE/TiO$_{2-x}$ (bottom electrode) interface shows Ohmic contact.

When a metal, such as Pt, contact with a semiconductor, such as TiO$_{2-x}$, a Schottky barrier is formed at the metal-semiconductor interface.[10] Moreover, Schottky barrier closely relates to the defect state at interface, and it has been reported that Schottky barrier could be collapsed by interface states with high density, such as high concentration of V_O.[11] Due to the low oxygen partial pressure during reactive sputtering process, the TiO$_{2-x}$ layer contains large amount of V_O, which makes the Pt BE/ TiO$_{2-x}$ interface an Ohmic contact. On the other hand, the V_O

concentration is locally lowered at the TiO_{2-x} /Pt TE interface because the oxygen partial pressure in air is much higher than that during the reactive sputtering of TiO_{2-x} layer. The assumed V_O profile and the equivalent circuit are shown in Fig. 3(a).

(2) Intermediate state: After apply a pulse voltage with mediate height, a certain amount of V_O locals at both Pt BE/TiO_{2-x} and TiO_{2-x}/Pt TE interface, resulting them exhibit Ohmic-like behaviors, as shown in Fig. 3(b). This state may correspond to the I-V curve after -5.0/900ms as shown in Fig. 1(a).

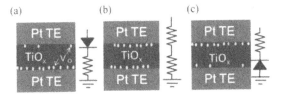

Fig.3 V_O profile and Equivalent circuit of Pt/TiO_{2-x}/Pt: initial state (a), intermediate state (b), switched state (c).

(3) Switched state: After applying a pulse voltage of enough height and width, V_O at TiO_{2-x} /Pt TE interface becomes many, while it becomes few at TiO_{2-x} /Pt BE interface, results in the Ohmic contact at TiO_{2-x} /Pt TE and Schottky barrier at TiO_{2-x} /Pt BE, as shown in Fig.3 (c).

Relationship between V_{switch} and T and w:
(1) With increase of T, V_{switch} decreases: Higher T results in higher diffusion coefficient of V_O, indicating drift of V_O under a certain pulse bias becomes easy. Therefore, V_O profile transform between initial state (Fig. 3a) and switched state (Fig. 3c) could be obtained at low pulse height.
(2) With increase of w, V_{switch} decreases: Longer w contributes more V_O drift from TE to BE even at lower pulse height, responding the results that switch could be obtained at lower V_{switch} if w becomes longer.

CONCLUSIONS

Current-voltage (I-V) characteristics were investigated in Pt/TiO_{2-x}/Pt devices. In the initial state, a rectifying property has been observed. After applying a pulse bias, the polarity of the device switch to an opposite direction, suggesting promising functionality of the engineered TiO_{2-x}/Pt interface. Influence of operation temperature and pulse bias width on this switch characteristic has been studied, and it exhibits that the rectifying polarity switch in the Pt/TiO_{2-x}/Pt is attributed to a thermal and dynamic dependence process. Therefore, local drift of oxygen vacancy (V_O) due to the applying pulse voltage is considered as the most possible origin.

REFERENCES
[1] D. B. Strukov, G. S. Snider, D. R. Stewart and R. S. Williams, Nature **453**, 80(2008).
[2] J. J. Yang, D. Pickett, X. M. Li, D. A. A. Ohiberg, D. R. Stewart and R. S. Williams, Nature nanotechnology 3, 429(2008).
[3] H. Shima, N. Zhong and H. Akinaga, Appl. Phys. Lett. **94**, 082905(2009).
[4] N. Zhong, H. Shima and H. Akinaga, Jpn. J. Appl. Phys., (in press).

[5] D. C. Cronemeyer, Phys. Rev. **87**, 876(1952).

[6] H. Miyaoka, G. Mizutani. H. Sano, M. Omote, K. Nakatsuji, and F. Komori, Solid State Commum. **123**, 399(2002).

[7] M. Janousch, G. I. Meijer, U. Straub. B. Delley, S. F. Karg, and B. P. Andreasson, Adv. Mater. (Weinheim. Germ.) **19**, 2232(2007)

[8] G. I. Meijer, U. Staub, M. Janousch, S. L. Johnson, B. Delley, and T. Neisius. Phys. Rev. B **72**, 155102(2005).

[9] J. R. Jameson, Y. Fukuzumi, Z. Wang, P. griffin, K. Tsunoda, G. I. Meijer, and Y. Nishi, Appl. Phys. Lett. 91, 112101(2007).

[10] W. Schottky, Naturwissenschaftern **26**, 843 (1938).

[11] S. M. Sze *Physics of Semiconductor Devices* 2nd ed. (Wiley, Hoboken, 1981).

Mater. Res. Soc. Symp. Proc. Vol. 1174 © 2009 Materials Research Society   1174-V06-35

Semiconductor-Metal Phase Transition in Doped Ion Beam Synthesized VO$_2$ Nanoclusters

H. Karl, J. Dreher and B. Stritzker
Institut für Physik, Universität Augsburg, D-86135 Augsburg, Germany

ABSTRACT

We have synthesized W and Mo doped VO$_2$ nanoclusters embedded in 200 nm thick thermally grown SiO$_2$ on 4-inch silicon wafers by sequential ion implantation of the elements V, W, Mo and O. The implantation energies have been chosen to locate the mean projected range in the centre of the SiO$_2$ thin film. A post implantation rapid thermal annealing step in flowing Ar at 1000°C for 10 min leads to the growth of doped VO$_2$ nanoclusters. The optical properties of the nanoclusters were analyzed by temperature dependent spectral ellipsometry in the spectral range of 320 to 1700 nm. It will be shown, that the semiconductor-metal phase transition hysteresis width starting at 50K in the undoped case can be systematically closed by increasing dopand concentration.

INTRODUCTION

Some of the vanadium oxides show a semiconductor-metal transition (i.e. VO$_2$ at 68°C or V$_2$O$_3$ at -123°C) [1,2] for that reason they attract increasing interest both in scientific research and technological applications. The near room temperature transition temperature of VO$_2$ is very attractive for many technological applications. The phase transition is marked by a strong decrease in electrical resistivity and an important change in optical transmittance and reflectance in the near infrared spectral region. Due to their functionality these materials might find applications in infrared optical modulators and switches [3,4], smart windows, waveguides and adaptable photonic crystals. With the semiconductor-metal transition the material undergoes also a structural phase transition (e.g. VO$_2$ transforms from a monoclinic in the semiconducting to a tetragonal structure in the metallic state). In particular the semiconductor-metal phase transition temperature of VO$_2$ can be altered and adjusted by doping making it possible to adapt that material to specific application requirements.

Due to the structural phase transition single crystals and thin films experience damage when cycled through the phase transition. This disadvantageous effect is largely reduced for material with nano crystalline morphology, where the energetic conditions prevent the generation of crystal defects. Apart these practical considerations nano-crystallinity gives rise to new physical phenomena. So it has been found, that nanoscopic grain size causes hysteresis of the semiconductor-metal phase transition and changes the transition temperature [5]. Most of the experiments have been performed on nanocrystalline thin films where the vanadium oxide crystallites are attached to each other forming grain boundaries. There is only a limited number experiments on well separated nanocrystallites embedded in an electrically insulating and optically transparent matrix. Those nano-composite thin films allow it to study effects of the localization of the phase transformation to a nanoscopic volume. This is particularly interesting since some of the properties are governed by the interdependent semiconducting and metallic domain growth.

EXPERIMENT

In this work the embedded surface-near nanocrystalline VO_2 precipitates have been synthesized by implantation of V^+ ions with energy of 100 keV and O^+ ions with 36 keV at room temperature in stoichiometric fluence ratio [6]. The dopands W and Mo have been co-implanted with energies of 330 keV and 165 keV. These implantation energies result in projected ranges of 90 nm and thus overlapping concentration profiles well centred in 200 nm thick thermally grown SiO_2 for all implanted ion sorts. A post-implantation rapid thermal annealing step at 1000°C for 10 min in Ar at atmospheric pressure leads to the formation of doped VO_2 nanocrystallites. X-ray diffraction and temperature dependent Raman studies revealed that the nanocrystals are composed of VO_2.

The implantations have been performed in 4 inch wafers by using a combinatorial materials synthesis method, which allows the generation of samples with different dopand concentrations synthesized under otherwise identical preparation conditions. This was realized by a specially designed implanter target end station with two moveable shields in front of the wafer. They cover up stepwise parts of the wafer surface as a function of the fluence to be implanted. In this way a lateral pattern of distinct fluence ratio combinations of dopand to V and O atoms is obtained. Technical details of the applied apparatus are described in [8,9,10]. In figure 1 elemental mappings of V, O and Si of cross sectional transmission electron microscopy (TEM) of an undoped sample are shown. The VO_2 nanocrystals are embedded in the centre of the SiO_2 layer and there is a sharp segregation of the elements detectable. There is a clearly separated bimodal particle size distribution, with a layer of particles with diameters smaller than 10 nm in a distance of approximately 20 nm from the SiO_2/Si interface and a layer of large particles with a diameter of approximately 100nm. The volume fraction of the small particle fraction can be neglected. It can thus be assumed, that the small particle ensemble will not determine the measured semiconductor-metal transition properties. Moreover the particles contain material voids clearly visible in the V and O map.

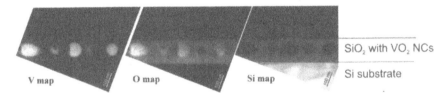

Figure 1. Elemental TEM mapping of V, O and Si.

For all samples the fluence of V and O was kept constant at $8x10^{16}$ at. cm^{-2} and $16x10^{16}$ at. cm^{-2}. The current of the co-implanted W and Mo ion beam was controlled. In order to quantify the dopand concentration and composition of the nanocrystalline VO_2-SiO_2 composite RBS measurements and simulations of the spectra have been performed. The result is shown in table 1. The (W+Mo)-to-V atom number ratio increases from 3 to 9 %. Mass separation of $^{118}W^{2+}$ and $^{59}Mo^+$ was not achievable due to their nearly equal charge-to-mass ratio. The W-to-Mo ratio increased with operating time of the implanter ion source arc-chamber. The (V+Si)-to-

O ratio of the V containing layer portion (also listed in table 1) coincides well with the stoichiometric element ratio of the compounds VO_2 and SiO_2, respectively.

Table 1. Atom number ratios determined from RBS measurements.

RBS atom number ratios					
(W+Mo)-to-V ratio	**0.031**	**0.050**	**0.065**	**0.079**	**0.093**
W-to-Mo ratio	0.198	0.264	0.422	0.586	0.723
(V+Si)-to-O ratio	0.621	0.505	0.532	0.542	0.504
V-to-Si ratio	0.398	0.393	0.394	0.391	0.375

The phase transition has been probed optically by using a spectral ellipsometer equipped with a peltier heating and cooling stage for thermal cycling from 0 to 100°C. The wavelength dependent rotation of the polarization and the phase shift has been measured after reflection of the probe beam as a function of substrate temperature. The semiconductor-metal transition can be detected most sensitively by the rotation of the polarization. Representative raw spectra of the polarization are shown in figure 2 were the polarization angle is plotted in the wavelength range between 320 and 1700 nm for heating starting in the semiconducting state at 25°C to 94°C were the material is fully metallic. There are also two additional spectra shown (figure 2) which have been measured at temperatures (75 and 77°C) well in transition regime of the semiconductor-metal phase transition. With increasing temperature there is a clear change of the polarization angle for wavelengths above 600 nm.

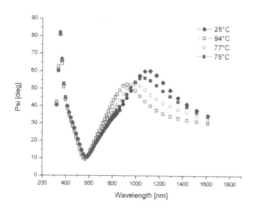

Figure 2. Polarization angle versus wavelength measured at different temperatures during heating of the sample.

The hysteresis loop has been recorded at a wavelength of 1200 nm, the results are plotted in figure 3 a) to f). The measurements have been performed by stepwise temperature increase (heating sequence) well above the upper semiconductor-metal transition temperature so that the material is fully metallic. The cooling temperature sequence was done likewise except for sample of figure 3 f) where the fully semiconducting state is reached only at temperatures below 0°C.

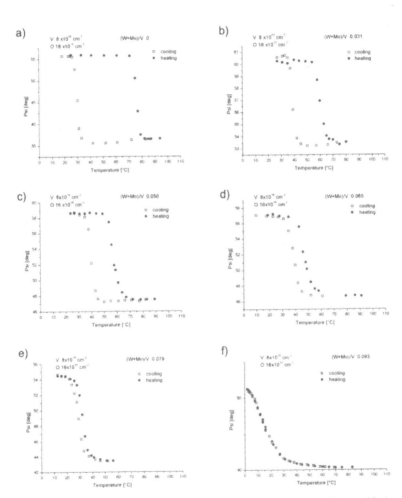

Figure 3. Hysteresis loops for different dopand concentrations corresponding to table 1.

DISCUSSION

This study shows that a single layer of electrically isolated nanocrystals of VO_2 can be synthesized in a thin layer of thermally grown SiO_2 by sequential ion implantation followed by a short thermal annealing process step. The nanocrystallites formed show a nearly spherical shape, most probably due to the short growth time. The characteristic temperatures of the semiconductor-metal phase transition hysteresis loop are summarized in figure 4. For undoped the sample (figure 4 a)) the upper transition temperature lays above and the transition temperature for the cooling sequence lays well below that of bulk VO_2 material. This gives rise to a loop width of approximately 50°C which only can be observed in nanocrystalline VO_2 samples. With increasing dopand concentration the hysteresis loop closes (figure 4 b) – e)). At a (W+Mo)-to-V atom number ratio of 0.093 the loop is completely closed (figure 4 f)). In parallel the critical temperature declines. For the closed loop case the determination of this temperature was skipped due the broad phase transition regime reaching well below 0°C.

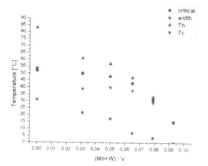

Figure 4. Characteristic temperatures of the hysteresis loops of figure 3. The width represents the difference between the transition temperature of the heating sequence (T_h) and that of the cooling sequence (T_c). The critical temperature (critical) is the median value between T_h and T_c.

Bulk VO_2 or thin films show only a very narrow hysteresis and in most reported cases the loop width is smaller than 5°C. It has been shown that the preparation technique of nanocrystalline VO_2 material has an effect on the hysteresis and that its width increases with decreasing crystallite size. Despite these findings and extensive studies the origin of the hysteresis remains still an open question. The largest hysteresis loop widths have been found in samples with isolated VO_2 nanocrystals. For thermodynamic reasons nanocrystals have only a small defect density compared to bulk or thin film material. It can thus be assumed, that a semiconducting or metallic domain completely fills an individual VO_2 nanocrystals. Interaction between the domains is strongly suppressed due to separation by the dielectric SiO_2 in between the crystallites. When nucleation sites are missing superheating or undercooling effects become effective in the majority of the nanocrystals of the ensemble. When introducing dopand atoms into the nanocrystals the number of nucleation sites will increase and lead to a reduction of the maximum and minimum temperature of the hysteresis [7]. In parallel the transition temperature decreases for the dopands of this study.

CONCLUSIONS

It was shown that stoichiometric implantation of V and O into SiO_2 followed by a rapid thermal annealing step results in the formation of an embedded VO_2 nanocrystal layer. The nanocrystals show a clear semiconductor-metal phase transition. The loop width of its hysteresis is significantly larger than in single crystals or thin films, which suggests that the nanocrystallites are filled with either single semiconducting or metallic domains. Doping with W and Mo introduces defects which act as nucleation sites and thus reducing the domain stability. At a certain dopand concentration the hysteresis loop closes. Moreover, increasing the dopand concentration also results in a reduction of the transition temperature and a widened phase transition regime. In that way the fully semiconducting state can be shifted well below room temperature.

In further experiments, we will determine more detailed dopand dependencies similar to figure 4 by doping Mo only. It can be expected that the hysteresis loop can be tailored without changing its critical temperature by co-implanting transition metals which shift the transition temperature in opposite directions.

ACKNOWLEDGMENTS

The authors thank J. Lindner for kindly performing the TEM elemental mappings.

REFERENCES

1. F. J. Morin, Phys. Rev. Lett. **3**, 34 (1959).
2. M. Gupta, A. J. Freeman, E. E. Ellis, Phys. Rev. B **16**, 3338 (1977).
3. R. Lopez, L. A. Boatner, T. E. Haynes, Appl. Phys. Lett. **85**, 8, 1410–1412 (2004).
4. Changhong Chen, Renfan Wang Lang Shang Chongfeng Guo, Appl. Phys. Lett. **93**, 171101 (2008).
5. R. Lopez, L. A. Boatner, T. E. Haynes, Apll. Phys. Lett. **79**, 19, 3161–3163 (2001).
6. R. Lopez, L. A. Boatner, T. E. Haynes, L. C. Feldman, , R. F. Haglund, J. Appl. Phys., **92**, 4031 (2002).
7. E. U. Donev, R. Lopez, L. C. Feldman, R. F. Haglund, Nano Letters **9**, 702–706 (2009).
8. H. Karl, I. Grosshans, B. Stritzker, Meas. Sci. & Technol. **216**, 396–401 (2004).
9. H. Karl, "Combinatorial Ion Beam Synthesis of II–VI Compound Semiconductor Nanoclusters", *Combinatorial and High-Throughput Discovery and Optimization of Catalysts and Materials,* ed. by R. Potyrailo and W. F. Maier (CRC-Book, 2006).
10. I. Grosshans, H. Karl, B. Stritzker, Mat. Sci. and Eng. B-Solid State Mat. for Adv. Technol. **101**, 1-3, 212–251 (2003).

Mater. Res. Soc. Symp. Proc. Vol. 1174 © 2009 Materials Research Society   1174-V06-11

Investigation of MgAlON Films on Electron Emission Properties

Mikihiko Nishitani[1,2], Masahiro Sakai[2], Masaharu Terauchi[1] ,Yukihiro Morita[1] and Yasushi Yamauchi[3]

[1]Co-operation Laboratory of Panasonic, Osaka University, Suita, Osaka, Japan

[2]Image Devices Development Center, Panasonic Corporation, Moriguchi, Osaka, Japan

[3]National Institute of Material Science (NIMS), Tsukuba, Ibaraki, Japan

ABSTRACT

The film of MgO-AlN system was formed by the conventional magnetron sputtering process. It is observed that a Fermi level could be raised by letting AlN composition increase, confirmed by ultraviolet (UV) photon energy dependence of the electronic emission, and investigated with XPS that the surface reaction, such as surface hydroxide / carbonate of MgO, could be controlled in addition of AlN. The glow discharge characteristics for MgAlON films were evaluated from the minimum pressure that a glow discharge is started under constant RF power, which correspond to the result of the analysis for Metastable De-excitation Spectroscopy (MDS).

INTRODUCTION

MgO is commercialized as a protective film of the plasma display with Ne/Xe discharge gas, since MgO is transparent, and sputtering tolerance is high, and the secondary electron emission coefficient (γ) for the Ne ion, which is necessary for the low voltage of the plasma discharge, is high. It will be more necessary to design the material with the high secondary electron emission coefficient (γ) for Xe ion to realize the plasma discharge which is low voltage with high efficiency in plasma display [1]. So far, it was suggested CaO, SrO, BaO or the composite materials of those [2]. However, due to the ionic bond of those materials, surface reaction of hydroxide / carbonate proceeds in the atmosphere quickly, and it is difficult to control the secondary electron emission coefficient (γ) . As a result, the oxide of the alkaline-earth metal does not reach practical use. On the other hand, we think that we may manufacture a film which has high secondary electron emission coefficient (γ) with surface stability, using MgO-AlN system. The system could be expected to result in occupied states near the of the valence band with the nitrogen incorporation compensated by electrons from Al of Mg site, similar to $TiO_{2-x}N_x$[3]. And the instability of the surface of MgO may be changed with the covalent bond of AlN. In this study, we try to design the material with MgO-AlN system to meet the demand of the plasma display which can expect high electron emission by the Xe ion irradiation with the surface stability.

EXPERIMENTAL DETAILS

The film of MgO-AlN system was manufactured by the conventional RF magnetron sputtering with MgO(99.99% purity), AlN(purity; 99.99%) sintering target of 2 inch and Al foil sheet (99.99% purity) and MgO granules(purity; 99.99%) for changing the film composition. The film deposition conditions were the following ; substrate temperature = 100°C, RF power = 50W, working pressure = 0.9 Pa using N_2 (purity; 99.99%) gas 100% except for the case of the MgO film deposition. The MgO film was deposited at the same pressure (0.9Pa) with Ar gas(purity; 99.9999%). A single crystal Si wafer with native oxide was used as a substrate for the film deposition. The film thickness was about 50 nm in all films. Additionally, the sputtering system used in this work was also set up to be able to turn RF power on to the substrate holder side for getting the discharge characteristic of the film. After the film deposition, the deposition chamber was pumped down again to less than 3×10^{-4} Pa immediately, and the pressure that the glow discharge begins was measured under the constant RF power to the substrate holder side. The pressure of Ar gas in that chamber was increased from the pressure of 0.05 Pa with 0.01 Pa/sec.

The film composition analyses and surface characterizations were carried out by X-ray Photoemission Spectroscopy obtained by Ulvac-phi system. The electron emission yield of the film was measured as a function of the incident photon energy with AC1 made by RIKEN KEIKI . The system of model AC-1 is an instrument for Photoelectron Yield Spectroscopy at atmospheric pressure that is an open counter equipped with an UV source. The open counter is a unique electron detector, which can be operated in open air, and has been used for photoelectron spectroscopy in the air. The MDS measurements at NIMS were carried out with meta-stable helium (He*) which have the energy of 19.8eV (2^3S). The incident He* beam yields electrons from the surface of the material due to the Auger neutralization process followed by the resonance ionization of He* or the Auger de-excitation process for 2S electron of He*. The MDS is an extremely surface-sensitive technique since only electrons at the outermost surface can contribute to the MDS spectra [4].

RESULTS AND DISCUSSION

Fig. 1 shows the compositions of the films in this work, which are plotted the N/(N+O) as a function of the Al/(Mg+Al). Those were analyzed from the measurement of XPS. As a result, a lot of content of the oxygen is included in the film since the aluminum atom is easy to take a lot of oxygen atoms compared with nitrogen atoms. There is a possibility that the films consist of MgO - AlN - Al_2O_3 co-ordinates. On the assumption of the idea, the lines, which corresponds to R=AlN/(Al_2O_3+AN)=100%, 75%, 45% and 28% respectively, are put on Fig.1. For example, the film on the line of R=45% consists of coordinates of the nitride of 45% with the oxide of 55% of Al.

The spectrum of C1s and O1s measured with XPS is shown in Fig. 2, picking up the sample along the line of R=28%. Since all samples are preserved in the package filled with nitrogen gas, they are exposed in the atmosphere for about 30 minutes till the measurement of XPS. The XPS spectrum of O1s and C1s reflects the hydroxide (denoted "OH-bond" in right side of Fig.2) , the carbon adsorption and the carbonate (denoted "C-C" and "Carbonate" in left side of Fig.2). The dotted lines in left side of Fig.2 show the results of the sputtering of about 2 nm from the surface with the argon ion of 500V acceleration. The O1s spectrum of the MgO film consists of two peaks that show the Mg-O bond and the Mg-OH bond, on the other hand, those spectra of the other films show single peak and shift from Mg-O bond (529eV) to Al-O bond (531eV) as the content of AlN increases. We speculate that the carbon adsorption and the surface hydroxide and carbonate was formed only at the film surface in the exposure to the atmosphere except the MgO thin film according to those results. The change in the surface reaction can be examined as a result of adding AlN to MgO in this study. As shown in Fig.2, we can observe to be able to control and suppress the formation of hydroxide / carbonate on MgO surface in addition of AlN or Al_2O_3. It is supposed that the surface reaction of the hydroxide and the carbonate was able to be controlled by the inclusion of Al-O and Al-N in the film, which are the covalent bond, compared with Mg-O bond.

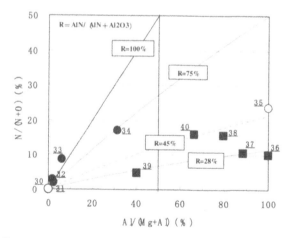

Figure 1 The film composition of N/(N+O) as a function of Al/(Mg+Al) with the calculated lines;R=AlN/(AlN+Al_2O_3)=100%,75%,45%,28%, respectively. The symbols of open circle, solid circle and solid square correspond to the film made by the sputtering process under Ar gas with the sintering target of MgO or AlN, under N_2 gas with Mg metal target and the sheet of Al, and under N_2 gas with the sintering target of AlN and the granule of MgO, respectively.

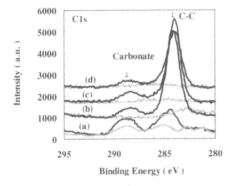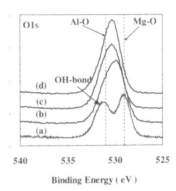

Figure 2 XPS spectra of MgAlON films deposited on Si substrates in this work. The sample notation (a) – (d) correspond to the film compositions following ; (a) MgO(No.30) , (b) Al/(Al+Mg)=0.40, N/(N+O)=0.05 (No.39), (c) Al/(Al+Mg)=0.89, N/(N+O)=0.11 (No.37) , (d) Al/(Al+Mg)=1.00, N/(N+O)=0.10 (No.36). Left side figure shows the spectra related with C1; solid lines show the samples exposed in the atmosphere for about 30 minutes and dotted lines show the samples after Ar ion sputtering of 2nm. Right side figure shows the spectra with O1s.

To evaluate whether it was possible to raise it in electronic energy of the valence band by the inclusion of AlN into MgO, the work function (Ef) was measured from the electron emission yield as a function of the UV photon energy. It was seen in Fig. 3 that the work function was raised with the increase of the nitrogen content in the film because of the existence of Al-N. As for the plasma discharge whose kinetic energy of the ion is less than 50 eV, the characteristics are mainly decided by the interaction of such low energy ion and a cathode material. In general, the break-down voltage and the pressure the glow discharge begins are decided depending on gas species and secondary electron emission coefficient (γ) of a cathode material through an Auger neutralizing mechanism. Because the first ionizing energy of the rare gas from He to Xe has about 10–25eV; He (24eV), Ar(16eV), Xe(12eV), it is necessary to study the surface electronic states in the region of about 25eV from the vacuum level on the cathode material. The high sensitive measurements around the valence band by XPS can be available with the Ulvac-phi XPS system (Quantera). In Fig.4, we show the result of the measurement with XPS from the vacuum level to about 25eV on the surface electronic states of the MgAlON films (the left side in figure). And we also show the result of the interaction with metastable state He (He*:2^3S, 19.8eV) at the outermost surface of MgAlON film, based on the resonance ionization (RI), the Auger neutralization (AN) and the Auger de-excitation (AD) process by using the MDS system

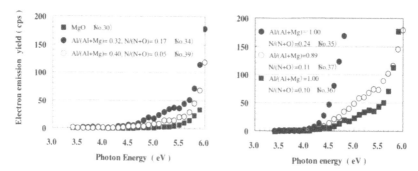

Figure 3   Electron emission yield as a function of UV photon energy. In left figure, the nitrogen content dependency in the center region of Al/(Mg+Al) is shown simultaneously with MgO of the reference sample (No.30). Right figure shows the effect of the nitrogen introduction in the region with little Mg.

that Yamauchi et al. developed [4] (the right side in figure).

The change of valence band shape and the energy shift of the O2s band are observed by the inclusion of Al and N into MgO from the result of XPS at the left of Fig.4. It is supposed that those changes are the results that the O-Al bond is introduced in addition of the O-Mg bond, confirmed by the first principle calculation named "SCAT" with the cluster of the Rock salt structure for MgO and the Corundum structure for Al_2O_3. The discrete variational (DV) - $X\alpha$ method named "SCAT" is one of the most useful techniques for approximately solving the Hartree-Fock-Slater molecular equation [5]. The shift of valence band to the low energy side and the tailing in the top of the valence band, on which the improvement of the secondary electron emission coefficient will be expect, cannot be observed from these data. In the right of Fig.4, the result of MDS is shown after the MgAlON films were processed in the ultra-high vacuum for 500℃ and 1 minute to remove hydroxide, carbon adsorption and carbonite on the surface. It seems that the electron emission characteristics follow the AD process. Because it is similar the shape of local DOS of O2p of the DV-$X\alpha$ calculation described before. Therefore, the spectrum observed with MDS is different from that of valence band (total DOS) observed from XPS. The spectrum shape doesn't qualitatively have the substantial change though the peak intensity and energy is changed with the introduction of AlN into MgO. However, it is suggested that the change in the cut-off energy around 0 eV is observed, which means the work function (Fermi energy) is changed. The electron DOS on the outermost surface measured from the vacuum level was analyzed, using the work function, φ, the cut-off energy, Ecut-off, Fermi energy level, Ef, of the MDS detection system which correspond to the kinetic energy of 14.7 eV and the He*

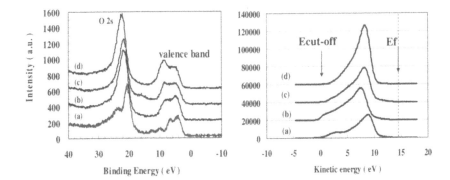

Figure 4 The valence band and O2s spectra of the various composition MgAlON films characterized by XPS (Left side) and the electronic density of states at the surface characterized by Metastable De-exicitatin Spectroscopy (Right side), using NIMS system [4] . The sample notation (a) – (d) correspond to the film compositions following ; (a) MgO(No.30) , (b) Al/(Al+Mg)=0.40, N/(N+O)=0.05 (No.39), (c) Al/(Al+Mg)=0.67, N/(N+O)=0.16 (No.40) , (d) Al/(Al+Mg)=1.00, N/(N+O)=0.10 (No.36).

energy, Eex, of 19.8 eV [6]. In other words, the work function, φ, of the films were estimated from the following equation;   Ef - Ecut-off = Eex - φ (1)
Right side of Figure 5 shows the density of state of the electron measured from the vacuum level based on the work function obtained from the analysis of MDS described in the above-mentioned. The original point of a horizontal axis is a vacuum energy level. Closeup figure around the threshold energy of electron emission is inserted in Figure 5. It is thought that these data correspond to the electron state on the extreme outermost surface compared with those of Figure 3 and left side of Figure 4. And the electron states obtained from MDS data are more closely related discharge phenomenon than those of Figure 3 and left side of Figure 4.

Assuming that the Auger neutralization is a predominant mechanism to cause the glow discharge (secondary electron emission), it is necessary for easy glow discharge that the electron should exist in the position of about half the energy of ionized energy for discharge gas measuring it from the vacuum level. Because the ionized energy of Ar is about 16eV, it is important for good discharge performance whether a lot of electrons with 8 eV or less from the vacuum level exist. Judging from the closeup in Figure 5, the glow discharge is expected to be caused easily than MgO(a) as for the sample of Al/(al+Mg)=0.40, N/(N+O)=0.05 (b). The left of Fig.5 shows the result that the pressure the glow discharge is begun was measured as a function of the RF input power, introducing the Ar gas to the deposition chamber, described in the

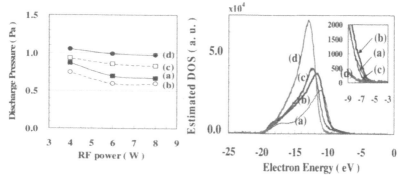

Figure 5 Comparison with the experimental discharge characteristics (left side) and the analyses of the density of state on the electron of the outermost surface from MDS date shown in Figure 4 (right side). The same sample notations (a) – (d) in Fig.4 are using ; (a) MgO(No.30) , (b) Al/(Al+Mg)=0.40, N/(N+O)=0.05 (No.39), (c) Al/(Al+Mg)=0.67, N/(N+O)=0.16 (No.40) , (d) Al/(Al+Mg)=1.00, N/(N+O)=0.10 (No.36).

experimental details. The left of Fig. 5 shows the result as is expected by the right of Fig.5.

CONCLUSIONS

We tried in this study to raise the energy of the top of the valence band by using composite materials of MgO and AlN and to obtain the surface stability. The former of the material design was able to be confirmed from the photon energy dependency of the emission of electron by the inclusion of AlN, especially N into MgO, and the latter of that was able to confirm those surface stabilities from the spectrum of C1s and O1s of XPS by the inclusion of AlN, especially Al.

It was able to be confirmed from the measurement of MDS that there is a composition in the MgAlON film with a larger secondary electron emission coefficient compared with MgO.

REFERENCES

1. G. Oversluizen, T. Dekker, M. F. Gillies, S.T. de Zwart, *SID '03 DIGEST, 28 (2003)*.
2. T. Shinoda, H. Uchiike, S. Andoh, *IEEE Trans.Electron Devices **ED-26**,1211 (1979)*
3. S.A. Chambers, S.H. Cheung, V. Shutthanandan, S. Thevuthasan, M.K. Bowman, A.G.Joly, *Chemical Physics 339, 27 (2007)*.
4. M. Kurahashi, Y. Yamauchi, *Surface Science **420**,259(1999)*.
5. H.Adachi, M.Tsukada, C. Satoko, *J. Phys. Soc. Jpn., **45**,875 (1978)*.
6. V.A. Bondzie, P. Kleban, D.J. Dwyer, *Surface Science **347**,319 (1996)*.
7. M.O. Aboelfotoh , J.A.Lorenzen , *J. Appl. Phys., **48 (11)** 4754 (1977)*

Applications

Mater. Res. Soc. Symp. Proc. Vol. 1174 © 2009 Materials Research Society   1174-V02-05

Integration of VO$_2$ Thin Films on Si (100) for Thermal Switching Devices Applications

Alok Gupta, Ravi Aggarwal, Jagdish Narayan
Department of Materials Science and Engineering, North Carolina State University, EB-I,
Centennial Campus, Raleigh, North Carolina 27695-7907

ABSTRACT

Thin films of vanadium dioxide (VO$_2$) exhibit an interesting semiconductor to metal transition (SMT) when heated above ~68^{0}C in which its resistivity changes by 3-4 orders of magnitude and its transmittance for IR wavelengths drops drastically. Integration of these thin films with Si (100) substrate is of immense technological importance due to its potential applications in sensor and memory based devices. Using pulsed laser deposition (PLD) we have demonstrated in this study that thin films of VO$_2$ can be grown epitaxially on Si (100) substrate using an intermediate tetragonal Yttrium-Stabilized Zirconia (YSZ) layer without any further annealing. X-ray diffraction (XRD) and cross-section transmission electron microscopy studies were performed on the films and they are found to be of highly epitaxial nature. Electrical resistivity measurement were carried out using the four-point probe method and SMT parameters were extracted using Gaussian fit of the data. The S-M transition parameters are in close proximity with parameters obtained from vanadium oxide films deposited on oxide based substrates such as Al$_2$O$_3$ or TiO$_2$.

INTRODUCTION

Ever since Morin [1] reported the reversible semiconductor to metallic phase transition in bulk single crystals of VO$_2$, there has been a considerable research interest in understanding the physics and structure property correlations associated with this transition [2-6]. This solid state phase transformation exhibits an abrupt resistivity and transmittance drop (in far infrared region) and therefore is of immense technological interest for their potential application in memory and sensor based devices [7-10]. One of the essential prerequisites for the realistic thermal switching based optoelectronic devices is to have methods that enable us to integrate the VO$_2$ thin films on Silicon substrate. In the past most of the research has been based upon using typically an oxide-based substrate such as Al$_2$O$_3$ and TiO$_2$; however, the fact that a specific substrate should be used substantially limits the fabrication of VO$_2$-based devices. Since Si (100) substrate is the choice for microelectronics applications, there is a thrust to process VO$_2$ thin films on Si (100) substrate. Though the epitaxial growth of VO$_2$ on Si (100) is anticipated via domain matching epitaxy (DME) technique [11], it is extremely difficult due to several technical issues. The direct deposition of VO$_2$ on Si (100) presents challenges problems such as formation of silicides or native amorphous silicon dioxide layer formation which would lead to the polycrystalline VO$_2$ thin films, diffused transition, and therefore reduced control over transition characteristics [12-14]. In this regard, the integration of thin films of VO$_2$ with silicon is of great technological importance from the viewpoint of multi-functionality on a chip as well as an expectation of prompt commercialization of thermal switching devices on the basis of well-established silicon technology.

In the present study, we have developed an approach based on pulsed laser deposition where we can grow epitaxial thin films of VO_2 on Si (100) using Yttrium-stabilized Zirconia (YSZ) as an intermediate layer for applications in Si based electronic and optical devices. X-ray diffraction and cross-sectional electron microscopy studies were performed to confirm the epitaxial nature of the thin films. Electrical resistivity measurements were performed using four point probe geometry and the results are presented.

EXPERIMENT

Thin film deposition was performed in a multitarget pulsed laser deposition chamber with a KrF excimer laser (Lambda Physik 210, $\lambda = 248$ nm). Schematic diagram of the set-up is shown in the figure 1(a). Commercially available YSZ target was used and VO_2 target was obtained by sintering of pressed VO_2 powder pellet (1" in diameter) at 1000 °C under Ar flow for 15 minutes. Si (100) substrates were cleaned ultrasonically in acetone following by similar cleaning in methanol. Substrates were then in the deposition chamber and the chamber was pumped down to a base pressure of 10^{-6} torr. Ceramic YSZ and VO_2 targets were ablated sequentially in the same run to grow YSZ buffer layers and VO_2 thin films on Si (100) substrates in order to prepare VO_2/YSZ/Si(100) heterostructures as shown in figure 1(b).

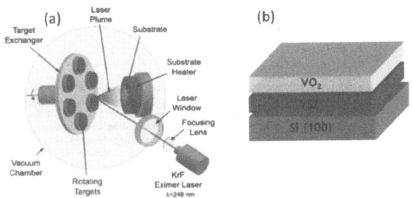

Figure 1. (a) Schematic of the pulsed laser deposition set-up and **(b)** sample architecture

A pulse width of 25 ns, pulse rate of 5 Hz, and an energy density of 2-3 J /cm^2 was used and the target-substrate distance was maintained at 4.5 cm during the depositions. Substrate temperature was maintained at 650 °C for deposition of both YSZ buffer layers and ZnO thin films. To remove the oxide on the silicon substrate, YSZ was deposited initially (1000 pulses) in vacuum. Oxygen was subsequently introduced into the chamber ($P_{O2} = 5 \times 10^{-4}$ Torr) for next 3000 pulses of YSZ. Substrate temperature was then lowered to 500 °C and the oxygen partial pressure was increased to 6-7 mtorr for VO$_2$ thin film deposition of 1500 pulses. An X-ray diffractometer with Cu Kα radiation (Rigaku, The Woodlands, TX, USA) was used to obtain x-ray diffraction data (θ-2θ scan). Cross-sectional samples of these films were characterized by a 2010F high resolution transmission electron microscope (JEOL, Tokyo, Japan) with Gatan image filtering. Four point probe measurement of electrical resistivity was done for temperatures ranging from room temperature to about 100 $^{\circ}$C at regular intervals using HP4155B semiconductor parameter analyzer.

DISCUSSION

In order to grow epitaxial thin films of vanadium dioxide (VO$_2$), YSZ was used as an intermediate layer. YSZ provides an epitaxial platform for integration with Si (100) and this approach has been developed by our group recently for integration of ZnO thin films [15]. Low thermal conductivity of YSZ also becomes useful for potential Si (100) based device integration because it can provide a good thermal barrier to reduce the effect of heating on other devices on the same Si substrate/chip.

Structural Characterization

Figure 2 shows the X-ray diffraction pattern (θ-2θ scan) of the VO$_2$/YSZ/Si (100) heterostructure in the as-deposited condition which suggests that YSZ grows in (002) orientation only. The pattern also shows that VO$_2$ thin film grows in (002) orientation on YSZ buffer layer and consists of a single phase.

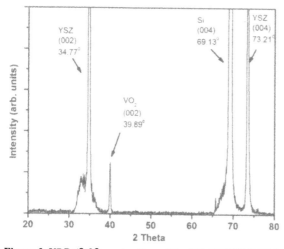

Figure 2. XRD (θ-2θ scan) pattern of the VO$_2$/YSZ/Si (100) heterostructure in as deposited condition.

Further analysis of the structure of the films was carried out using electron diffraction and high-resolution transmission electron microscopy (HRTEM). Figure 3(a) is a low magnification image of the as-deposited VO_2/YSZ/Si (100) heterostructure. From the cross-section TEM micrograph the thicknesses of the YSZ and VO_2 films were determined to be 100 nm and 55 nm respectively. A high resolution image of the VO_2/YSZ interface is shown in figure 3(b) which shows that there is no occurrence of interfacial reaction at the VO_2/YSZ interface.

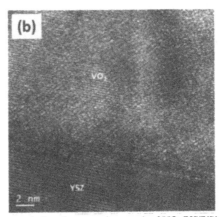

Figure 3. (a) Low-magnification cross-section transmission electron micrograph of VO_2/YSZ/Si (100) heterostructure **(b)** HRTEM image of VO_2/YSZ interface.

The structure was determined using the diffraction pattern taken in [110] zone of Si (100) substrate. Figure 4(a) shows the selected area electron diffraction for Si/YSZ. Upon indexing the pattern the epitaxial relationship it is found that the YSZ has tetragonal phase and the epitaxial relationship between YSZ and Si follows the condition such that [110] $_{YSZ}$ ‖ [100] $_{Si}$ and [001] $_{YSZ}$ ‖ [001] $_{Si}$. From the diffraction pattern of VO_2, shown in figure 4(b), the phase is identified to be VO_2 with a monoclinic structure and lattice parameters that are consistent with the values for bulk single crystal VO_2. It is derived from the diffraction pattern that VO_2 thin films grow on YSZ buffer such that [001] $_{VO2}$ ‖ [001] $_{YSZ}$ and that there is a possible ~6° rotation between [010] $_{VO2}$ and [110] $_{YSZ}$ which is presently being studied in detail.

Electrical Characterization

Resistance was measured using four point probe method and value of R/R_{max} (normalized) has been plotted as shown in figure 5(a). Analysis of the electrical resistivity data taken during the thermal cycling and calculation of different parameters associated with semiconductor to metal transition (SMT) using Gaussian fit model [16] is presented in figure 5(b). Hysteresis was found out to be ~ 6K and the order of magnitude change in the resistivity was about 3. For VO_2 thin films based on silicon substrate, these values are one of the best reported and are in close proximity of values reported for films deposited on single crystal oxide substrates such as Al_2O_3 (sapphire) [14, 17]. In context of potential integration, another important fact to note here in is that epitaxial YSZ buffer layer is dielectric in nature and therefore would reduce the leakage current. This is consistent with literature where it has been reported that YSZ when used as a gate electric can reduce the leakage current by about five orders of magnitude than that of SiO_2 for equivalent oxide thickness [18].

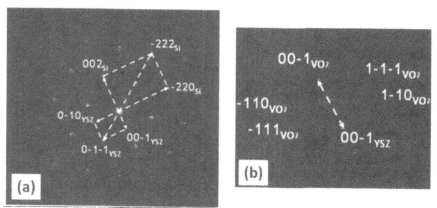

Figure 4. Selected area electron diffraction (SAED) patterns with sample titled along Si [110] zone axis (**a**) of Si/YSZ interface region and (**b**) of VO_2 film.

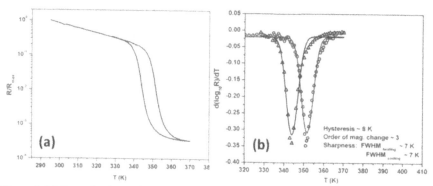

Figure 5. (**a**) Normalized electrical resistivity (R/Rmax) of epitaxial VO2 film as function of temperature. Lower curve is for cooling cycle while upper curve is for the heating cycle (b) The derivatives of log10R(T) for the heating (red, circle) and cooling (blue, triangle) curves are shown. Symbols are data points, and the lines are Gaussian fits whose minima and widths determine the transition temperature (TSMT) and SMT width (FWHM).

CONCLUSIONS

In summary, we have developed an approach using YSZ as an intermediate layer to grow epitaxial vanadium dioxide (VO_2) thin film on Si (100) substrate by employing pulsed laser deposition technique. This development is a big step in the direction of realization of VO_2 based thermally activated switching devices utilizing well-established silicon technology.

ACKNOWLEDGMENTS

Authors are pleased to acknowledge the financial support from National Science Foundation (NSF) under award No. 0803663.

REFERENCES

1. F. J. Morin, Phys. Rev. Lett. **3**(1), 34 (1959).
2. J. B. Goodenough, J. Solid. State. Chem. **3**, 490 (1971).
3. H. Liu, O. Vasquez, V. R. Santiago, L. Diaz, F. E. Fernandez, J. Elec. Mat. **33**(10), 1171 (2004).
4. S. Biermann, A. Poteryaev, A. I. Lichtenstein, and A. Georges, Phys. Rev. Lett. **94**, 026404 (2005).
5. J. Narayan and V. Bhosle, J. Appl. Phys. **100**, 103524 (2006).
6. M. M. Qazilbash, M. Brehm, Byung-Gyu Chae, P.-C. Ho, G. O. Andreev, Bong-Jun Kim, Sun Jin Yun, A. V. Balatsky, M. B. Maple, F. Keilmann, Hyun-Tak Kim, and D. N. Basov, Science **318**, 1750 (2007).
7. R. Kumar R. T., B. Karunagaran, D. Mangalaraj, S. K. Narayandass, P. Manoravi, M. Joseph, and V. Gopal, Sensors Actuators A **107**, 62 (2003).
8. W. R. Roach, Appl. Phys. Lett. **19**, 453 (1971).
9. G. V. Jorgenson and J. C. Lee, Sol. Energy Mater. **14**, 205 (1986).
10. J. S. Chivian, M. W. Scott, W. E. Case and N. J. Krasutsky, IEEE J. Quantum Electron. **21**, 383 (1985).
11. J. Narayan and B. C. Larson, J. Appl. Phys. **93**, 278 (2003).
12. R. Thompson, J. Appl. Phys. **52**, 6763 (1981).
13. K.N. Tu, J.F. Ziegler and C.J. Kircher, Appl. Phys. Lett. **23**, 493 (1973).
14. N. Yuan, J. Lia, G. Lib, and X. Chenc, Thin Solid Films **515**, 1275 (2006).
15. R. Aggarwal, C. Jin, P. Pant, J. Narayan, and R. J. Narayan, Appl. Phys. Lett. **93**, 251905 (2008).
16. D. Brassard, S. Fourmaux, M. Jean-Jacques, J. C. Kieffer, and M. A. El Khakani, Appl. Phys. Lett. **87**, 051910 (2005).
17. H. Jerominek, F. Picard, and D. Vincent, Optical Engg. **32**(9), 2092 (1993).
18. S. J. Wang, C. K. Ong, S. Y. Xu, P. Chen, W. C. Tjiu, A. C. H. Huan,W. J. Yoo, J. S. Lim, W. Feng and W. K. Choi, Semicond. Sci. Technol. **16**, L13 (2001).

Work Function Effects of Nano Structured ZnO Thin Film on the Acetone Gas Sensitivity

Seung Hyun Jee [1], Nitul Kakati [1], Soo Ho Kim [2], Dong-Joo Kim [3] and Young Soo Yoon [1,*]
[1] School of Advanced Material Engineering, Yonsei University, Seoul, Korea
[2] Department of Advanced Technology Fusion, Konkuk University, Seoul, Korea
[3] Materials Research and Education Center, Dept. of Mechanical Engineering, Auburn University, AL, USA

ABSTRACT

We deposited various nano structured ZnO thin films with plasma treatment on an alumina substrate and fabricate ZnO sensors for acetone gas detection. The ZnO sensors with various nano structures and the plasma treatment (PT) were deposited by radio frequency (RF) magnetron sputtering method with RuO_2 micro heater and Ru electrode. In order to control the work function intentionally, the various deposition conditions and the plasma treatment were used. Sensitivities of the ZnO sensor were measured in acetone vapor and air at 250℃. In conclusion, we suggested that not only plasma treatment can enhance the work function but also the sensitivity of ZnO sensors for acetone be improved on the work function increase of the plasma treated ZnO thin films with nano structures.

INTRODUCTION

Many researchers have recently focused on ZnO thin film because it is an inexpensive n-type and wide band gap (3.2 eV) semiconductor with high transmittance, good electrical conductivity and high gas sensing property.[1,2,3] These properties make the ZnO a multifunctional material that can be applied to sensing material of gas sensors and a transparent electrode of optical devices. [4,5]. Even though the ZnO thin film shows relatively high sensitivity for acetone gas, the operation temperature and detectable minimum acetone gas concentration are still high when we consider the real commercial application of the ZnO thin film based gas sensor. [6,7] The electron transfers on surface depends on work function [8]. However, influence of work function of ZnO thin film on acetone gas sensitivity has not been reported yet. In this study, the ZnO thin films with various deposition conditions and plasma treatment were fabricated by radio frequency (RF) magnetron sputtering on an alumina substrate with a RuO_2 micro heater and a Ru electrode. In order to control the work function intentionally, different deposition conditions and the plasma treatment were used. In this study, we investigated relations between sensitivity and the work function in the ZnO sensor for acetone with various work functions by various deposition conditions and the plasma treatment.

EXPERIMENT

The RF magnetron sputtering system with Zn target was used to deposit ZnO thin film on alumina substrate by reactive method. The sputtering gas was used to a mixture of $Ar-O_2$, and the rate of oxygen was varied from 10% to 50%. Thickness of ZnO thin films was 400nm. The chamber had a base pressure of 5×10^{-6} torr and a working pressure of 8.0×10^{-3} torr. Distances from target to substrate were 6.5 cm (in plasma) and 10 cm (out plasma). Table I shows the various deposition conditions of ZnO thin film by RF magnetron sputtering. After the deposition of ZnO thin films, the ZnO thin films with various deposition conditions were treated by oxygen plasma. Plasma treatments were carried out at the working pressure of 5.0 x 10^{-2} torr and the RF power of 50W for 10 minutes. After plasma treatment of ZnO thin films, patterned Ru electrodes were deposited on the sensing material by using RF magnetron sputtering system. Then, a patterned RuO_2 heater was also deposited on the back side of the substrate by RF magnetron sputtering. The work functions of ZnO thin films were measured by using Kelvin Probe. Besides, X-ray photoemission spectroscopy (XPS) and X-ray diffraction (XRD) were used to analyze the work function change. The sensitivities of the sensors for acetone concentration were calculated by measuring the resistances of the sensor in air (R_a) and in acetone gas (R_g) by using Keithley 2400 connected with a computer by a Labview program. The sensor sensitivities were measured at 250°C and concentrations of acetone vapor were 500 and 1500 ppm.

Table I. Various deposition condition of ZnO thin film by RF magnetron sputtering.

sample name	O_2(%)	power(W)	distance(cm)	Work function (eV)
S1	10% (Ar)36 : (O_2) 4	100	10	4 62
S2			6.5	4 75
S3		200	10	4 57
S4			6.5	4 75
S5	30% (Ar)28 : (O2) 12	100	10	4 55
S6			6.5	4 57
S7		200	10	4 55
S8			6.5	4 59
S9	50% (Ar)20 : (O2) 20	100	10	4 55
S10			6.5	4 69
S11		200	10	4 49
S12			6.5	4 66

DISCUSSION

The ZnO thin films deposited by various conditions had various work functions from 4.49 to 4.75 eV, as shown in table I. Error ranges of work functions of ZnO thin films measured by Kelvin probe was ±0.01eV. XRD spectra of the ZnO thin films were compared to analyze structural property of the ZnO thin films. Fig. 1 shows the XRD spectra of the ZnO thin films with various deposition conditions. The intensity of (002) peak is more in case of ZnO thin films deposited by using the RF power of 200W than the others. We cannot find the relation between the deposition condition (RF power and oxygen concentration) and work function of as deposited ZnO thin films. However, the ZnO thin films deposited with in-plasma condition had higher work functions than that with out-plasma condition. We expected that plasma treatment may affect the work function change of the ZnO thin films. Therefore, we investigated how plasma treatment affects the work function change.

130

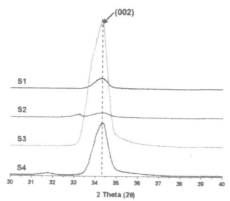

Fig. 1. XRD spectra of the ZnO thin films with various deposition conditions.

Fig. 2 shows the work function change of the ZnO thin films before and after plasma treatment and time-dependent work function change of ZnO thin films with various conditions. The ZnO films with plasma treatment had increased work function about 0.6 eV. We demonstrated that plasma treatment can increase the work function of ZnO thin films. Furthermore, we demonstrated that the ZnO thin film deposited with in-plasma condition has higher work function is because of plasma treatment effect when ZnO thin films were deposited. After plasma treatment, the work functions of ZnO thin films were from 4.8 to 5.55eV. The work functions of ZnO thin films were dramatically decreased after 60, 120, and 240 minutes.

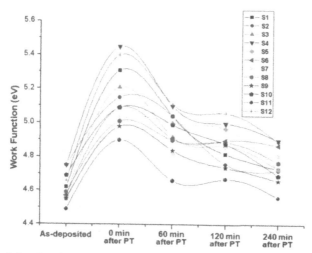

Fig. 2. Work function change of ZnO thin films before and after plasma treatment and time dependent work function change of ZnO thin films with various conditions.

The cause of the work function change was analyzed by using XPS spectra. We measured peak binding energy of oxygen atoms on ZnO surface by XPS. Fig. 3 shows the work functions of ZnO thin films and peak binding energies of the oxygen atom. As shown in Fig. 3, the work function and the peak binding energy of the oxygen atom was almost proportional. Besides, the peak binding energy of oxygen atom in ZnO films with plasma treatment was increased proportially. We could project the cause of work function change of ZnO thin films.

First, the plasma treatment removed dangling bond on ZnO surface due to bind the ZnO molecules on the ZnO surface and oxygen atoms by the plasma treatment. This means free electrons were removed or reduced from the ZnO surface. Therefore, the work function of the ZnO thin film with the plasma treatment was increased because energy to escape free electron was needed more.

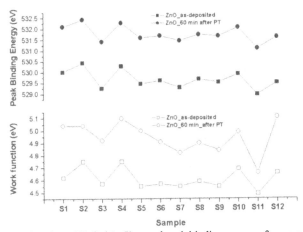

Fig. 3. Work function of ZnO thin films and peak binding energy of oxygen atom.

Second, the ZnO surface was oxidized by oxygen plasma. This means binding energy of oxygen atom can be increased as shown in Fig.3. Therefore, the bonding energy increase of oxygen atoms on ZnO surface restricts the emission of electrons and increases the work function. These results demonstrated that the bonding energy increase of oxygen atoms on ZnO with plasma treatment increases the work function of ZnO thin film. However, decrease of the plasma treatment effect depends on time. We need to development of time-independent surface modification in which work function was unchanged.

The sensitivity of the ZnO sensor for acetone is defined as $[(R_a-R_g)/R_a \times 100]$ where R_a is the electrical resistance of the sensor in air and R_g is the electrical resistance of the sensor in acetone-air mixture gas. The sensitivity of the ZnO sensor with the plasma treatment was more improved than those without the plasma treatment. These results mean electron swap property between ZnO surface and acetone gas was improved by plasma treatment.

In addition, the ZnO sensors by various deposition conditions without plasma treatment had high sensitivity difference, as shown in Fig. 4. However, The ZnO sensor with plasma treatment had not only higher sensitivity but also uniform sensitivity in various deposition conditions. Besides, the sensitivity difference of ZnO by plasma treatment was decreased. In Fig. 4, the sensitivity difference of as-deposited ZnO sensors with various conditions for acetone of 500ppm was 74.53%. However, the sensitivity difference of plasma treated ZnO sensors with various conditions for acetone of 500ppm was 47.85%. We can also see the sensitivity difference of ZnO sensor was decreased by plasma treatment in Fig. 4. In this result, we demonstrate the ZnO sensor with uniform sensitivity for acetone vapor can be fabricate by plasma treatment, even if ZnO thin films were deposited with various work functions by various deposition conditions.

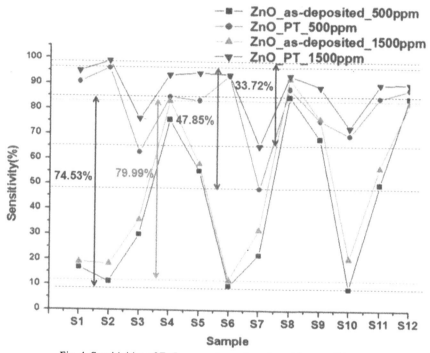

Fig. 4. Sensitivities of ZnO sensors by various deposition conditions.

CONCLUSIONS

The ZnO thin films with various work functions by RF magnetron sputtering were fabricated. Besides, the work functions of ZnO thin films were increased by RF oxygen plasma treatment. The cause of structural properties and work function increase of ZnO thin films were analyzed. The work function of ZnO thin films and binding energy of oxygen atoms on ZnO surface was almost proportional. In addition, the binding energy of oxygen atom on ZnO surface by plasma treatment was proportionally increased, comparing with the work function changes of the ZnO thin films. We expected that work function increase by plasma treatment is due to generate the dangling bond and increase the binding energy of oxygen atom by oxidation of the ZnO surface. The sensitivity of ZnO sensor for acetone was improved by plasma treatment. We demonstrated that electron swap property between ZnO surface and acetone gas was improved by plasma treatment. Therefore, we could fabricate the ZnO sensor with uniform and improved sensitivity for acetone by plasma treatment.

REFERENCES

1. R. L. Hoffman, B. J. Norris and J. F. Wager, Appl. Phys. Lett. 82, 733 (2003).
2. D. C. Look, Mater. Sci. Eng. B 80 (2001) 383.
3. A. Tsukazaki, M. Kubota, A. Ohtomo, T. Onuma, K. Ohtani, H. Ohno, S. F. Chichibu and M. Kawasaki, Jpn. J. Appl. Phys. 44, L634 (2005).
4. Amit Kumar Chawla, Davinder Kaur and Ramesh Chandra, Optical Materials. 29, 995 (2007).
5. E. Fortunato, P. Barquinha, A. Pimentel, A. Goncalves, A. Marques, L. Pereira and R. Martins, Thin Solid Films. 487, 205 (2005).
6. M. Suchea, S. Christoulakis, K. Moschovis, N. Katsarakis and G. Kiriakidis, Thin solid films, 515, 551 (2006).
7. A.M. Gas'kov, M.N. Rumyantseva, Russ. J. Appl. Chem. 74, 440 (2001).
8. K. H. Lee, H. W. Jang, K.-B. Kim, Y.-H. Tak, J.-L. Lee, J. Appl. Phys. 95, 586 (2004).

Mater. Res. Soc. Symp. Proc. Vol. 1174 © 2009 Materials Research Society   1174-V06-08

Application of Simple Cycloalkylsilanetriols as Surface Modifier for Inorganic Particles

Bok Ryul Yoo, Dong Euy Jung, and Joon Soo Han
Organosilicon Chemistry Lab, Korea Institute of Science & Technology

ABSTRACT

Simple cycloalkylsilanetriols RSi(OH)$_3$ [R $=c$- pentyl (1) and c-hexenyl (2)] were synthesized as white powders in 70 and 90% isolated yields, respectively, from the hydrolysis of the corresponding trialkoxysilanes with mild acidic water. Silanetriols 1,2 are soluble in water and polar organic solvents such as alcohol, acetone, THF, DMSO, and etc. They can be applied as surface modifiers for inorganic materials such as silica and titania through simple two step processes: 1) organosilanetriol interact with hydroxyl groups on inorganic materials through hydrogen bonding to make molecular layered thin coating on the inorganic surface, 2) then can undergo condensation by heating about 110 ℃ above to form M-O-Si covalent bond to give hydrophobic silica. With the treatment of ca. 50 nm spherical silica with 5 wt % alkylsilaneol 1, no precipitation of silica particles in water was observed, indicating good surface modifiers for silica particle. In this presentation we will discuss the application of silanetriols as surface modifier for inorganic materials such as silica and titania particles.

INTRODUCTION

Inorganic oxide particles such as silica and titania are utilized as fillers for inorganic-organic polymer composites to obtain excellent fill-ability, thermal conductivity, viscosity characteristics, good mechanical, and electric properties.[1] There are hydrophilic characters due to hydroxyl groups on surfaces of inorganic particles. This hydrophilic surface does not process good compatibility with organic polymer, and therefore the silica cannot be dispersed very well into the polymers. While the inorganic particles with hydrophilic surface easily adhere to each other through hydrogen bonding network leading to irregular agglomerations.[2] The agglomerations of the inorganic particles can form a network through the whole polymer matrix and occlude liquid polymer in their inter-particle voids, thereby affecting the rheology of the composite and giving a significant rise to the viscosity as inorganic filler loading increases.[3] Thus the surface modification of inorganic particles can be changed to be hydrophobic property from hydrophilic. Generally silylation of inorganic particles using organosilane coupling agents such as organoalkoxysilane and organochlorosilane is the most commonly used method for surface modification.[4,5] The organic groups were attached to the surface of inorganic particle via the non-hydrolysable Si-C covalent bond and functioned as hydrophobic or/and network binder. It has also been known that the stability of particulate SiO$_2$ or TiO$_2$ dispersed in non-polar organic media is significantly improved when the surface of the particles is modified with organosilane coupling agents. There are many factors affecting the result of inorganic particle surface treatment, including the type of coupling agent, concentration of coupling agent, treatment time and predisperse method. Since these factors might interact with each other in determining the final result of the surface treatment, the motivation of this work is to investigate the optimal condition for inorganic particle surface treatment in order to obtain hydrophobic particles. In the surface modification of inorganic particles, silanol (Si-OH)[6,7] formed by the hydrolysis of silane coupling agents plays a important role to make a Si-O-M covalent bond to the inorganic surface.[2] Such siliane coupling agents requires moisture in their hydrolysis step and then eventually volatiles such as alcohol or hydrogen halide are evolved as volatile organic compounds. Herein we wish to report a facile surface modification of inorganic oxide particles using relatively stable organosilanetriols instead of existing organoalkoxysilane coupling agents.

EXPERIMENTAL

General comments

Inorganic nanoparticles such as silica (SiO_2, 50 nm average diameter, Rodia) and titania (TiO_2, 100 nm average diameter, Aldrich) were commercially available and used as received. Cyclic alkenes such as c-pentene and c-hexene, and methanol were purchased from Aldrich. Trichlorosilane from Gelest, Inc. The cyclic alkyltrichlorosilanes[8,9] were prepared by the hydrosilylation of trichlorosilane with cycloalkenes and reacted with methanol and sodium methoxide to give cyclic alkyltrimethoxysilanes.

Characterization of surface inorganic particles

The morphology of particles including size, shape, and agglomeration of nano silica and titania were characterized by a Scanning Electron Microscope (SEM: Hitachi S-3000H). To obtain SEM images, a drop of silica nanopatricles suspension was placed on a scanning electron microscopy (SEM) specimen holder. The sample was allowed to dry under ambient temperature for 2 h prior to imaging. Then, the sample was Pt-Pd sputtered for 60 s to minimize charging. The surface chemistry of the inorganic particles was studied using Diffuse Reflectance Fourier Transform Infrared (DR/FTIR) spectrometer and Mercury-Cadmium-Telluride (MCT) as a detector.

Surface modification of inorganic particle

In a 250 mL round bottom flask, 50 g of silica particle (115 m²/g, ZEOSIL® 115GR, Rodia) was dispersed with stirring in 100 mL of methanol. A solution of 1.0 g of cyclopentenylsilanetriol dissolved in 5 mL methanol was added into the flask and stirred for10 min. Some volatile compounds including methanol and water were removed from the slurry phase of silica and methanol using a rotary evaporator. Then the remaining particles were dried at 120 ℃ for 6 h in an oven. In order to examine the morphology of the silica particles before and after treatment, the dried silica particles were analyzed with a SEM technique. Their SEM images are shown in figure 1. Also, the surface-treated silica was analyzed by infrared spectrometry using DRIFT. According to the same procedure described in nanosilica treatment with silanetriol above, treatment of titanium dioxide (TiO_2: Aldrich, 99.8%)) surface with silanetriol **1** was carried out. 10 g of titanium dioxide was treated with 1.0 g of **1**. The morphologies of titanium dioxide particles before and after surface treatment were characterized by SEM, and the observation results are shown in figure 3. The modification of nano titania were also studied using characterization techniques including FTIR. The tests for floating of inorganic particles on water and for dispersion in toluene solvent were conducted, respectively.

Cycloalkylsilanetriols RSi(OH)$_3$ [R =c- pentyl (1) and c-hexenyl (2)] were synthesized as white powders in 70 and 90% isolated yields, respectively, from the hydrolysis of the corresponding trialkoxysilanes with mild acidic water adjusted to around pH 3-4 by formic acid. Silanetriols 1, 2 are soluble in water and polar organic solvents such as alcohol, acetone, THF, DMSO, etc., and were used as coupling agents for surface modification of inorganic particles. A mixture of de-ionized water and methanol was used as the medium for inorganic particles treatment, and the formic acid was used to adjust the pH value of the medium. Such cycloalkylsilanetriols were applied to surface modifiers for inorganic particles. At first, the surface modification of silica particles were conducted using organosilanetriols instead of existing organoalkoxysilane coupling agents in previously reported. This modification process using organosilanetriols is simple and easier than that using typical alkoxysilane coupling agents: 1) a slurry mixture of silica and organosilanetriol in methanol was admixed at room temperature for 10 min, 2) Then volatiles such as water and methanol were removed from the slurry suspension of silica and methanol at temperatures ranging to 60 ℃ from room temperature using a rotary evaporator, and the remaining silica particles were dried in a oven at 120 ℃ for 6 h. It was effective at high temperature of 110 ℃ and above for the condensation of the silanol groups between silica and organosilanetriol affording a Si-O-Si covalent bond formation. The silanol of organosilanetriol can either react with the silanol groups on the surface of silica to form a stable functionalization on the silica surface, or with neighboring other silanetriol molecules to form siloxane by further polymerization.[7b] Such silanetriols are so relatively stable in neutral medium at room temperature even though they undergo condensation reaction under thermal and acidic/basic conditions to afford a variety of polymeric silsesquioxanes. They can attach effectively to silanol of silica surface through hydrogen bonding and their self-condensation can be reduced under controlled condition. Such silanetriol-coated silica formed by the hydrogen bonding between the silanol of silica with silanetriol can be converted to covalent bond of siloxane (Si-O-Si) by baking at the temperature of 110 ℃ and above. When the treatment of silica particle was carried out by varying the amounts (1-10 wt %) of silanetriol 1 to investigate the optimum amount of 1 for surface modification, the precipitation rate of modified silica in water medium decreased as the amount of the silanetriol used increased. With the treatment of ca. 50 nm spherical silica (BET 115m^2/g) with 5 wt % (or above) alkylsilaneol 1, no precipitation of silica particles in water was observed, indicating a good surface modifier for silica particle. There is no problem involving organic volatile compounds (especially alcohol) in composite materials with organic polymer because of use of organosilanetriolinstead of alkoxysilane coupling agent. The functional groups on the surface of silica treated with silanetriol have been characterized by the analysis of IR data. But the functional groups at the surface of silica particles are not easy to detect due to the strong and broad bands of Si-O (1020-1250 cm^{-1}) and – OH (3300-3700 cm^{-1}).[10] Especially it was well-known that the strong band of OH (3300-3700 cm^{-1}) attributed to the isolated silanol unit can be observed in the case of raw silica nanoparticles, while the amount of isolated silanol unit decreased by modifying the silica surface with silanetriol coupling agent. In a FTIR spectrum, the intensity of the absorption bands between 3300 and 3700 cm^{-1} related to the presence of silanol group (Si-OH) was lower in the treated silica than that in the untreated silica. In a surface analysis of surface of inorganic particles, the data of DR/FTIR spectra can give more significant information on the functional groups at the surface. a DR/FTIR spectrum of silica modified by silanetriol 1, characteristic absorption bands were observed at 2929 cm-1 (asym, C-H stretching), 2857 cm-1 (sym, C-H stretching) and 1457 cm-1 (C-H bend), negative broad bands appearing between 1150 cm-1 and 1050 cm-1 were considered to be attributable to Si_s–O–Si.[11] From the FTIR analyses, it can be ascertained that the functional groups are successfully introduced onto the nanosilica surface. Silica modified by this method was used for further characterization. The surface morphology of silica particles was examined by the analysis of SEM data. The SEM images of silica particles before and after surface treatment with 7 wt % silanetriol **1** are showed in figure 1. As shown in figure 1, the morphology of silanetriol-treated silica was almost equal to the individual particle size and shape of silica provided by manufacturer when compared with untreated silica, suggesting an agglomeration did not take placed during surface treatment of particles with organosilanetriol.

Figure 1. SEM Images of Untreated Silica (left) and Silanetirol-Treacted Silica (right)

The modification properties of silica particles were also examined for their dispersibility into water and into nonploar organic medium (toluene). The precipitation rate of silica particles before and after surface treatment were studied in water medium and compared with each other after they were shaken in water and then left to stand for 1 h. The images of silica particles before and after surface treatment were are shown in figure 2. As can be seen from the photographs in figure. 2, the silica particles before surface treatment was precipitated down within a few minutes, while the silica particles after surface treatment floated on the water for 1 h and even until 1 d, indicating they were modified well to be hydrophobic silica. The surface properties of nanosilica modified with other cyclic organosilanetriol **2** were similar to that with cyclopentylsilanetriol **1**. But it was not obvious whether the difference of the surface properties between two silicas modified silanetriols **1** and **2** occurred at the present.

Figure 2 . Photographs of Untreated of Silica Particles (left) and Silanetriol-Treated Silica (right) after standing for 1 h.

Generally the functionality of the organic moiety of silanetriols has played an important role in an application of modified silica to the hybridization with organic polymers. They are readily dispersed in non-polar organic media and the resulting dispersions are considerably more stable and resistant to phase separation than untreatedsilica particles. Now the hybridization of nano silica with organic polymers (synthetic rubbers) are in progress.

According to the same procedure described in silica treatment with silanetriol 1 above, treatment of titanium dioxide surface with 5 wt % silanetriol 1 was carried out. The modification properties of titania particles were also examined for their dispersibility into water and into nonploar organic medium (toluene). With the treatment of ca. 100 nm spherical titania with 5 wt alkylsilaneol 1, no precipitation of silica particles in water was observed, indicating a good surface modifier for titania particle. The morphologies of surface of titanium dioxide particles before and after surface treatment with 5 wt % silanetriol 1 were examined by a SEM technique, and their SEM images are in figure 3. The SEM pictures in figure 3 also shows that the morphology of titania particles modified with silanetriol 1 hardly had any effect in the particle agglomerations when compared with untreated titania particles suggesting that the silanetriol is very excellent modifier in term of titania modification is related on the hydrophilicity of their fhnctional groups. The characterization of titania modified with silanetriol was conducted using the same methods described in that of silica modified with the same silanetriol. The functional groups on the surface of titania modified with silanetriol can be identified well by IR technique.

Figure 3. SEM Images of Untreated Titania (left) and Silanetirol-Treacted Titania (right)

This process is simple and easy for surface modification of inorganic oxide particles and can achieve effectively inorganic particles to be hydrophobic through mixing, evaporation, and baking steps. Now the hybridization of titania with organic polymers (synthetic rubbers) are in progress.

CONCLUSIONS

This work has demonstrated that cyclic organosilanetriols are useful surface modifiers for the inorganic particles. The results showed that the inorganic particle treatment by organosilanetriols is simple and easy process through three steps 1) a mixing of inorganic particles and organosilanetriols in solvent (water or methanol), 2) removal of volatiles by evaporation, 3) a simple baking at 110 ℃ and above in oven. The results showed that the surface morphologies of inorganic particles before and after silanetriol treatment were not significantly changed in their SEM images. The test of dispersibility of modified inorganic particles in water medium showed good hydrophobic characteristics. They are readily dispersed in non-polar organic media (toluene) and the resulting dispersions are considerably more stable and resistant to phase separation than untreated inorganic particles. The research will be helpful to elucidate the modification of inorganic oxide particles with alkoxysilane coupling agent.

ACKNOWLEDGMENTS

This research was supported by a grant from the Fundamental R&D Program for Core Technology of Materials funded by the Ministry of Knowledge Economy and partially by Korea Institute of Science and Technology.

REFERENCES

1. Z. Hua, W. Shishan, and S. Jian, *Chem. Rev.* **108**, 3893 (2008).
2. D. C. Bradley, R. C. Mehrotra, and D. P. Gaur, *Metal Alkoxides*, Academic Press: New York, 1978
3. (a) J.-M. Yeh and K.-C. Chang, *J. Ind. Eng. Chem.* **14**, 275 (2008). (b) S.-K.Yoon, B.-S. Byun, S. Lee, S. H. Choi, *J. Ind. Eng. Chem.* **14**, 417 (2008). (c) S.-S. Choi and J.-C. Kim, *J. Ind. Eng. Chem.* **13**, 950 (2007).
4. A. V. Rao, M. M. Kulkarni, D. P. Amalnerkar, and T. Seth, *Appl. Surf. Sci.* **206**, 262 (2003).
5. A. Krysztalfkiewicz and B. Rager, *Colloid & Polym. Sci.* **266**, 1435 (1988).
6. P. D. Lickiss, *Adv. Inorg. Chem.* **42**, 147 (1995).
7. (a) J. H. Kim, J. S. Han, M. E. Lee, D. H. Moon, M. S. Lah, and B. R. Yoo, *J. Organomet. Chem.* **690**, 4677 (2005). (b) J. H. Kim, J. S. Han, W. C. Lim, and B. R. Yoo, *J. Ind. Eng. Chem.* **13**, 480 (2007).
8. B. Marcinec,in *Hydrosilylation: A Comprehensive Review on Recent Advances,* ed. J. Matisons (Springer, 2009).
9. Z. V. Belyakova, E. A. Chernyshev, P. A. Storozhenko, S. P. Knyazev, G. N. Turkel'taub, E. V. Parshina, and A. V. Kisin, *Russian J. General Chem.* **76**, 925 (2006).
10. A. L.Smith, *The Analytical Chemistry of Silicones* (Wiley, 1991).
11. M. L. Hair, and C.P. Tripp, *Langmuir* **7**, 923 (1991).

Mater. Res. Soc. Symp. Proc. Vol. 1174 © 2009 Materials Research Society   1174-V06-12

Development of a Sensor for Polypropylene Degradation Products

Shawn M. Dirk,[1] Patricia S. Sawyer,[1] Robert Bernstein,[1] James M. Hochrein,[2] Cody M. Washburn,[1] Stephen W. Howell,[3] and Darin C. Graf[4]

[1] Organic Materials Department, [2] Materials Reliability Department, [3] Rad Hard CMOS Technology Department, [4] Advanced Sensor Technologies Department, Sandia National Laboratories, P.O. Box 5800, Albuquerque, NM 87185

ABSTRACT

This paper presents the development of a sensor to detect the oxidative and radiation induced degradation of polypropylene. Recently we have examined the use of crosslinked assemblies of nanoparticles as a chemiresistor-type sensor for the degradation products. We have developed a simple method that uses a siloxane matrix to fabricate a chemiresistor-type sensor that minimizes the swelling transduction mechanism while optimizing the change in dielectric response. These sensors were exposed with the use of a gas chromatography system to three previously identified polypropylene degradation products including 4-methyl-2-pentanone, acetone, and 2-pentanone. The limits of detection 210 ppb for 4-methy-2-pentanone, 575 ppb for 2-pentanone, and the LoD was unable to be determined for acetone due to incomplete separation from the carbon disulfide carrier.

INTRODUCTION

The aging of partially crystalline olefins, such as polypropylene, is particularly complex. Polypropylene has been exposed to thermal, radiation, and thermal combined with radiation accelerated aging environments. Isotopically labeled carbons (13C) have been selectively positioned at chemically distinct locations in the polymer and studies performed on materials containing enrichment at each position. Selective labeling, combined with 13C nuclear magnetic resonance (NMR) and mass spectroscopy, has provided the ability to follow each chemically distinct carbon throughout the degradation process.[1, 2] Some of these degradation products including 4-methyl-2-pentanone, acetone, and 2-pentanone were selected as targets for a newly developed chemiresistor sensor system that uses a siloxane matrix.[3]

As an example, Scheme 1 shows the proposed degradation pathway of polypropylene resulting in 4-methyl-2-pentanone. Since isotopically labeled polypropylene was used with mass spectroscopy characterization, each of the three distinct atoms was followed with unprecedented detail. This resulted in the proposed mechanisms as well as extreme confidence in the assignment of the degradation products. [2]

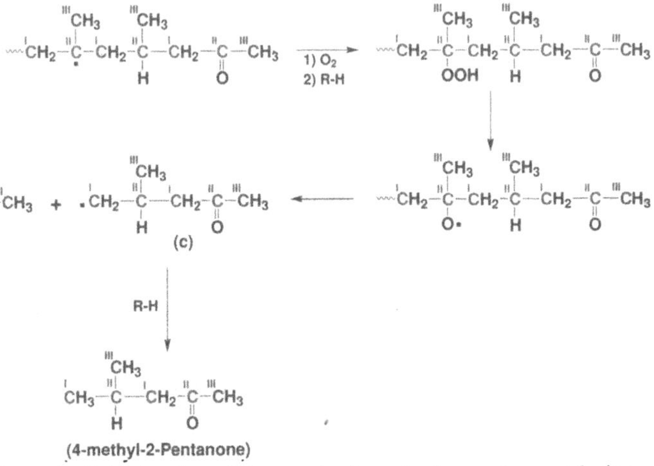

Scheme 1. Oxidative degradation mechanism of polypropylene producing 4-methyl-2-pentanone.[2]

Chemiresistors may be a potentially useful tool to monitor polymer breakdown as a function of time. Chemiresistors are very low power sensor devices comprised of a chemoselective film that is applied to a set of inderdigitated electrodes (IDEs). The sensors are probed as a bias is applied to one of the IDEs. Several publications have described the signal transduction mechanism as an activated tunneling model that contains two terms as shown in equation (1).[4-6]

$$\sigma = \sigma_o e^{(-\delta\beta)} e^{(-Ea/RT)} \qquad (1)$$

In this expression, σ is the electronic conductivity of the film, δ is the interparticle distance, β is the electronic coupling coefficient (effectively a measure of the density of states (DOS) available between conducting particles), Ea is the activation energy for electron transfer between adjacent conducting particles, R is the gas constant, and T is the absolute temperature. The first exponential factor takes into account the effect of nanoparticle spacing and the DOS overlap between nanoparticles, and the second exponential term is related to the permittivity of the film.

In most cases when chemiresistors are exposed to an analyte the analyte affects both factors of the activated tunneling model serving both as a solvent swelling the nanoparticle/matrix as well as changing the effective dielectric constant of the sensing film. In the case of low dielectric constant analytes, the first exponential factor dominates and conductivity decreases. In the case of analytes with high dielectric constants the second factor dominates, and conductivity increases in the presence of the analyte. All the polypropylene degradation analytes chosen contain a ketone functional group which is known to increase the dielectric constant.

EXPERIMENTAL DETAILS

IDEs were fabricated from Au with a Ti adhesion layer on quartz substrates. The gap between the IDEs was 5 μm. Prior to sensor chemistry deposition the IDEs were cleaned with acetone (30 s bath and rinse) and rinsed with DI water. The device was then submerged in Piranha (1:1 30% hydrogen peroxide and conc. sulfuric acid) for 2 min and rinsed with copious amounts of DI water. CAUTION: PIRANHA IS A VERY STRONG OXIDIZER AND REACTS VIOLENTLY WITH ORGANICS. The surface was dried with CO_2.

The chemoselective sensors were fabricated using an aqueous solution of gold nanoparticles that were then crosslinked in the presence of the silica precursor, tetraethyl orthosilicate (TEOS), with a conjugated α,ω-dithiolate. The dithiolate was prepared from the *in-situ* deprotection of an acetate capped thiol terminated phenylene ethynylene molecule. The crosslinked nanoparticles and silica matrix were drop-cast onto IDEs as previously reported.[3]

Sensors were exposed to 4-methyl-2-pentanone, acetone, and 2-pentanone. The analytes are dissolved in 1 mL CS2 and injected into an Agilent 6890 split injection gas chromatograph equipped with 1 meter 100 μm ID capillary column coated with a polydimethylsiloxane (PDMS) stationary phase at ambient temperature. The carrier gas was hydrogen and the split was 30. The flow rate through the column was 30 sccm. The injection port was heated to 250 °C. Gas flow exited the capillary column and entered a custom made test fixture illustrated in Figure 1. Gas left the test fixture and returned to the HP FID detector heated to 250 °C. The injection volume was 1 μL and repeated injections were performed with an auto-injection tower for reproducibility.

Figure 1. Illustration of test fixture and actual test fixture used for evaluation of the sensor coating chemistry.

After Au nanoparticle/silica assembly, the functionalized substrate was placed in an evaluation fixture. The lid of the test fixture contained pogo pins that connected to a Keithley 6487 current pre-amp voltage source. The lid included a milled channel to allow gas flow over the 5 μm IDEs. The lid was sealed with an external O-ring. The voltage source and current measurement instruments were controlled by a custom LabVIEW program. During an analyte exposure, 5 V was typically applied to the substrate and the current was measured as a function of time.

RESULTS AND DISCUSSION

As the sensor was exposed to the analytes the current increased as compared with conventional chemiresistors where the current decreases. Figure 2 shows the response of the chemiresistor sensor upon exposure to a mixture of carbon disulfide carrier and 4-methyl-2-pentanone. The 4-methyl-2-pentanone was delivered to the sensor via a gas chromatography system configured using a 1 m PDMS coated column. The short section of gas chromatography column was used to separate the carbon disulfide carrier from the 4-methyl-2-pentanone. The concentrations of 4-methyl-2-pentanone in carbon disulfide were varied to determine the limit of detection (LoD) of the coated sensor. Sensors prepared using the siloxane matrix were very stable and showed little degradation as a function of time.

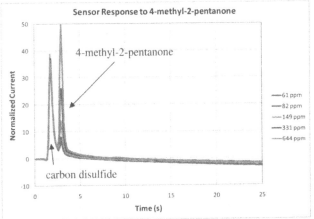

Figure 2. Electrical response of a chemoselective sensor film exposed to several concentrations of 4-methyl-2-pentanone.

Figure 3 shows the LoD estimate for the sensor upon exposure to 4-methyl-2-pentanone. The LoD number was determined at 3 times the noise of the sensor. The LoD was determined to be 210 ppb for the test fixture and the sensor upon exposure to 4-methyl-2-pentanone.

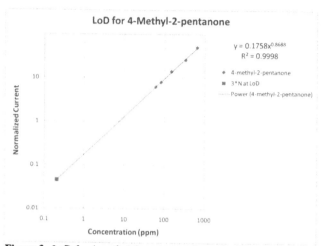

Figure 3. LoD for 4-methyl-2-pentanone determined by extrapolation

Figure 4 shows the response of the sensor when exposed to five solutions of 2-pentanone in carbon disulfide. The solution concentrations were varied from 54 ppm to 586 ppm.

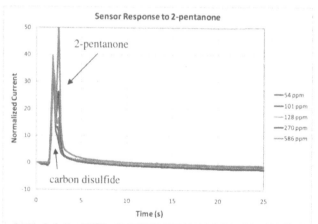

Figure 4. Electrical response of a chemoselective sensor film exposed to several concentrations of 2-pentanone.

The normalized response to each analyte concentration was plotted and a line was fit to the data and extrapolation was used to determine the LoD at a normalized current reading of three times the noise was calculated to be 575 ppb and is shown in Figure 5.

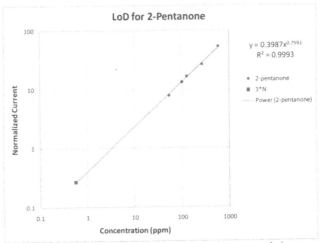

Figure 5. LoD for 2-pentanone determined by extrapolation

Sensors were also exposed to acetone. Presumably a longer column was needed to evaluate acetone as it co-eluted with the carbon disulfide carrier. This co-elution can be seen can in the sensor response to acetone plot (Figure 6). In the previous two analyte cases the sensors produced normalized current reading of ~40, where as the solutions of acetone produced a response of 50 for the carbon disulfide peak indicating that acetone was present as well.

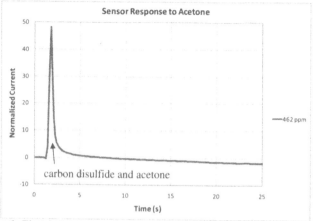

Figure 6. Electrical response of a chemoselective sensor film exposed to several concentrations of acetone.

CONCLUSIONS

The chemiresistor sensor system comprised of Au nanoparticles crosslinked with a bisthiol terminated phenylene ethynylene in a siloxane matrix cast onto IDEs was evaluated as a possible sensor for degradation products of polypropylene. The analyte was delivered to the sensors via a gas chromatography system equipted with a 1 m PDMS coated GC column. Sensors prepared showed low limits of detection for 4-methyl-2-pentanone and 2-pentanone and were extremely robust showing little change in response over a few months.

ACKNOWLEDGEMENTS

Sandia is a multiprogram laboratory operated by Sandia Corporation, a Lockheed Martin company, for the United States Department of Energy under contract DE-AC04-94AL8500).

Mater. Res. Soc. Symp. Proc. Vol. 1174 © 2009 Materials Research Society   1174-V09-36

Optimizing the Screen-Grid Field Effect Transistor for High Drive Current and Low Miller Capacitance

Y. Shadrokh[1], K. Fobelets[1], J. E. Velazquez-Perez[2]

[1]Imperial College of London, Department of Electrical and Electronic Engineering, Exhibition Road, London SW7 2BT, UK

[2]Universidad de Salamanca, Departmento de Física Aplicada, Edificio Trilingüe, Pza de la Merced s/n, E-37008 Salamanca, Spain

ABSTRACT

Reduction of parasitic capacitances and improvement of the on-off current ratio (I_{ON}/I_{OFF}) can be achieved by increasing the gate control in Field Effect Transistors (FETs). Multiple gated FETs (MugFETs) lend themselves well for this. The MugFET investigated in this manuscript is the Screen Grid FET (SGrFET) that consists of multiple gate cylinders inside the channel perpendicular to the current flow. In this work we illustrate, using 2D Technology Computer Aided Design (TCAD), that the multiple geometrical degrees of freedom of the SGrFET can be exploited to simultaneously optimise the on-current, I_{ON} and the gate-drain Miller parasitic capacitance for increased switching speed.

INTRODUCTION

The Screen Grid Field Effect Transistor (SGrFET) is an alternative multi-gated FET (MugFET) [1]. The SGrFET is defined on SOI (Silicon-On-Insulator) substrates and has a non-traditional gating geometry, illustrated in fig. 1. The gate consists of multiple cylindrical cavities with a thin thermal oxide sidewall and a poly-Si or metal filling. These gate cylinders (fingers) are standing perpendicular to the current flow in the SOI body (channel). Different gate cavity configurations are possible [1]. Highly doped source and drain areas are located at both sides of the device and have the same width as the device, avoiding contacting problems to small areas. For optimum performance the channel doping is low to un-doped in order to preserve high mobility values and is of the same doping type as the contact regions, unlike in traditional MOSFETs [2]. The device operation is essentially MESFET-like (Metal Semiconductor FET) – the channel width is determined by the extension of the depletion between two gate cylinders in the same row. The role of the second row of gate fingers (near the drain) is to effectively control short channel effects. The SGrFET outperforms other MOSFET structures in the sub-threshold and weak inversion regions, because in these regimes the carriers have higher mobilities. The multiple geometrical degrees of freedom of the SGrFETs reduce the parasitic gate-drain capacitance and improve the sub-threshold slope. These make the SGrFET suitable for digital applications with improved switching speed and improved gate control over the channel. In this paper all the simulations are carried out using a 2D Technology Computer Aided Design (TCAD) device simulator, Medici[TM] [3]. In the first section we show how changing the device structure improves rise and fall times. The second section shows how to improve I_{ON} and I_{OFF} in P-type devices for CMOS circuits. The technique improves the rise time in the output voltage at the same time. These results show the potential of SGrFETs for digital applications.

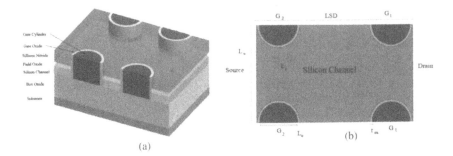

Figure 1. (a) Schematic 3D view of one unit cell of a two-gate row SGrFET. The layer structure and the gate cylinders perpendicular to the surface inside the Si body are shown. Source and drain are resp. the left and right side of the figure. (b) Cross section through the channel parallel to the top surface. Definition of geometrical parameters: L_u: width of the unit cell, L_c: width of the channel, L_0: diameter of the gate, t_{ox}: gate oxide thickness and L_{SD}: source-drain distance. G1 is the source-side gate, G2 the drain-side gate.

MILLER EFFECT AS A FUNCTION OF DEVICE GEOMETRY

In this section some key geometrical parameters, from the multiple available in the SGrFET, are used to improve the switching speed of the device in a CMOS circuit. Up till now only symmetric gate cylinder configurations were studied. We will demonstrate in this paper that the SGrFET not only shows robust characteristics when used in an asymmetric device structure but that it also improves overshoot/undershoot in the output voltage of the CMOS [4]. Fig. 2 shows four gate configurations of the SGrFET. All of the devices have a source to drain distance of L_{SD}=140 nm and an actual device width of L_u = 74nm. In L1, all the gate diameters (L_0=30 nm) and oxide thicknesses are the same (t_{ox}=2nm). In L2, all the gates have the same oxide thickness, but gate row G1 has L_0=15 nm and gate row G2 has L_0=30 nm. In L3, all cylinders have L_0=30 nm, but the gate row G1 has an oxide thickness two times that of gate row G2. In L4, all the gate diameters and oxide thicknesses are the same, but gate row G1 is shifted 10 nm towards gate row G2.

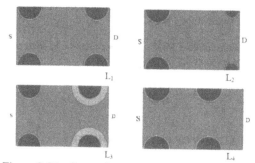

Figure 2. L1: all gate diameters and oxide thicknesses are the same with L_0=30nm and t_{ox}=2nm. L2: all oxide thicknesses are the same as in L1 but, L_{0G1}=30nm and L_{0G2}=15nm. L3: all cylinders have L_0=30nm, t_{oxG1}=t_{oxG2}/2=2nm. L4: all gate diameters and oxide thicknesses are the same as in L1 but G1 shifts 10 nm towards G2.

The circuit mode simulation tool of Medici is used to model the CMOS circuit using the four different device geometries shown in fig. 2. The result of the transient analysis is shown in fig. 3 a & b when the input signal goes from 1V to 0V and from 0V to 1V respectively.

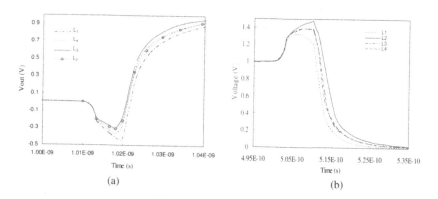

Figure 3. Output voltage of the SGrFET CMOS for the different device structures given in fig. 2 (a) Rise time, (b) Fall time.

The results show that increasing the distance between the drain ohmic contact area and the drain side gate cylinder G2 or/and decreasing the diameter of the drain side gate cylinders or/and increasing the oxide thickness of the drain side gate cylinder, all improve the overshoot and undershoot of the output voltage in the CMOS circuit considerably. This is because these geometrical changes influence the parasitic Miller capacitance of the device. At the same time there is no degradation of the rise and fall times. Table I summarises the maximum over/undershoot as a function of device geometry in the CMOS circuit. The results in table I show that the CMOS circuit based on L3 has the smallest over/undershoot values thus significantly improving the overall switching speed of the CMOS.

Table I. Over/Undershoot of the output voltage of the CMOS circuit as a function of device geometry given in fig. 2.

	L1	L2	L3	L4
Overshoot (V)	1.48	1.38	1.31	1.38
Undershoot (V)	-0.45	-0.37	-0.3	-0.38

I_{ON} AND I_{OFF} IMPROVEMENT AS A FUNCTION OF DEVICE GEOMETRY

It has been shown in [1] that I_{ON} increases by a factor N (N=1,2,3..) through the addition of N unit cells to the device given in fig. 1. This can be done without influencing the other device characteristics. This technique was used to improve I_{ON} and switching speed in SGrFET logic gates [5]. Here we show (see fig. 4) that an increase in the gate diameter, without changing L_c, also leads to a significant improvement in I_{ON} and I_{OFF}. This increase in the I_{ON} is a result of a reduction in the current path from the drain to the source and the reduction in I_{OFF} is due to an increase of the gate cylinder over the channel area. There is a slight increase in the threshold voltage (4mV) and improvement in sub-threshold slope while DIBL remains constant (22.2 mV/V) when the gate cylinder diameter (L_0) increases from 30nm to 60nm. This is simply due to an increase in the control of the gate over the channel without increasing the actual width of the device unlike in the MOSFET. Table II summarizes the I_{ON}/I_{OFF} ratio as a function of the gate diameter for P-type SGrFET.

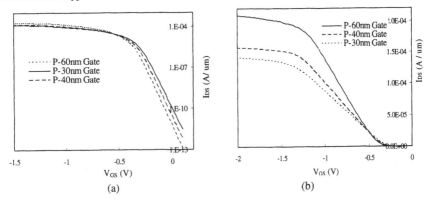

(a)   (b)

Figure 4. P-type SGrFET DC transfer characteristics show I_{ON} and I_{OFF} as a function of gate diameter (a) I_{OFF} (log scale), (b) I_{ON} (linear scale). V_{DS}=1V

As shown in fig. 4, an increase in the gate diameter improves ON and OFF current. However, this comes at the cost of increasing gate-drain capacitance which leads to an increase in over/undershoot in the output voltage of an inverter. However, this increase in gate-drain capacitance can be resolved by using one of the four techniques explained in section 1. Therefore, one can take advantage of the larger I_{ON}/I_{OFF} ratio without an increase in the actual width of the device and at the same time reduce the over/under shoot in output voltage.

The DC threshold voltage V_{th}, sub-threshold slope S and DIBL of a symmetric gate N-type SGrFET (see L_1 in fig. 2) with $L_0 = 30$ nm and $t_{ox} = 2$ nm and of an asymmetric gate P-type SGrFET (see L_4 in fig. 2) with $L_0 = 40$ nm are summarised in Table II. The distance between G1 and the drain area is 20nm and $t_{ox} = 2$ nm. These results show that the relocation or enlargement of gate cylinders does not have any impact on the threshold voltage. The sub-threshold slope improves as the gate cylinder distance both vertically and horizontally in the channel reduces or the diameter of the gates increases. This is due to an increase of the gate control over the channel. It is important to point out that increasing the distance between G_1 and the drain area does not have any effect on the gate control over the channel. This is due to the radial gating action of the gate cylinders which can still provide the control over the distance between G_1 and the drain area. Shifting G_1 towards the source area increases the gate-source capacitance which helps in reducing the overshoots and undershoots in the output voltage obtained of the CMOS in transient analysis.

Table II. DC parameter extraction

	V_{th} (V)	S (μm^2)	DIBL (mV/V)
Symmetric N-type	0.38	63	55.5
Symmetric P-type	-0.3	63	22.2
Asymmetric P-type	-0.32	61	22.2

We use an asymmetric device with a larger gate diameter for the P-type FET and a symmetric device with smaller gate diameter for the N-type device in the CMOS circuit to increase the level of the current in the P-type device. It should be noted that, unlike the ordinary MOSFET, an increase in the level of the current in the P-type device does not require an increase in device width. This is due to the difference between the operational concept of the MOSFET and SGrFET which are based on the inversion layer and the degree of the depletion area respectively.

CMOS CIRCUIT OPTIMIZATION

Fig. 5 shows two CMOS circuits, symmetric and asymmetric, based on the two different P-type device structures. In the asymmetric CMOS circuit the gate cylinders in the P-type device increase to improve the I_{ON} and I_{OFF} current without changing the width of the P-type device while Miller capacitance reduces using the flexible device geometry by increasing the distance between the drain-side gate cylinder and the drain area ohmic contact (L3 device configuration in fig. 2).

153

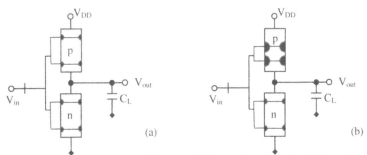

Figure 5. (a) Symmetric CMOS circuit with L_0 = 15 nm for both N- and P-type FET. (b) Asymmetric CMOS circuit with L_0 = 15 nm for the N-type and L_0 = 20 nm for P-type FET where the drain side gate row is shifted towards the source over a distance of 20nm.

Fig. 6 shows that the gate-drain capacitance as a function of input voltage for the two CMOS circuits of fig. 5. In the P-type region of fig. 6, the line with square markers gives the Miller capacitance for the symmetric CMOS structure and the dashed line corresponds to the asymmetric structure. It is clear that the gate-drain capacitance reduces considerably when the P-type device is switching in the asymmetric CMOS structure. The N-type devices are symmetric in the both CMOS circuits thus no change in this region will occur. A significant improvement in the overall Miller capacitance can be achieved if the asymmetric P-type device is used in the CMOS circuit.

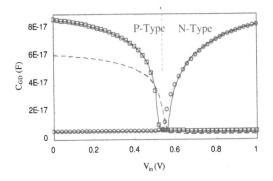

Figure 6. Gate-drain capacitance of the N-type and P-type devices as a function of input voltage in the CMOS circuit. The line with square markers is for the symmetric CMOS circuit. The dashed line is for the asymmetric CMOS circuit. N-Type: C_{GD} when the N-type device switches; P-Type: C_{GD} when P-type device switches.

Fig. 7 gives the results of the transient analysis of the output voltage for symmetric and asymmetric CMOS circuits. There is a considerable reduction in undershoot and overshoot in the output voltage of the asymmetric CMOS circuit. As explained before, the significant reduction in over/undershoot is due to the reduction in C_{GD} in the P-type device, while fall time improves through an increase in gate cylinder diameters of the P-type device. Also rise/fall times improve considerably as a result of an increase in the gate diameter of the P-type device to improve the I_{ON}.

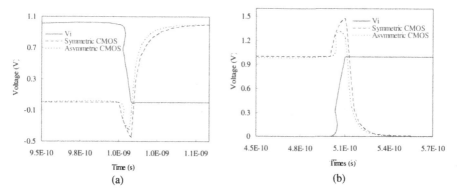

Figure 7. Transient waveforms of the symmetric and asymmetric CMOS circuits (a) Rise time (b) fall time.

Table III summarises the rise and fall times as a function of device geometry in the CMOS circuit shown in fig. 5. The rise time improves to 16ps while the fall time remains constant. The over/undershoot improves significantly by, respectively, 11% and 22% for the asymmetric CMOS circuit. The improvement of the asymmetric CMOS circuit results in a reduction of the delay time and power dissipation.

Table III. Rise and fall time and over/undershoot for the two symmetric and asymmetric CMOS circuits given in fig. 5.

	Symmetric CMOS	Asymmetric CMOS
Rise Time (ps)	23	16
Fall Time (ps)	10	10
Overshoot (V)	1.49	1.32
Undershoot (V)	-0.45	-0.35

CONCLUSION

We have shown that the multiple geometrical parameters of the SGrFET lead to different logic circuit configuration schemes to improve performance. The gate-drain capacitance can be reduced when the device's drain-side gate row diameter or oxide thickness reduces. These changes reduce the parasitic Miller effect considerably in the transient analysis of CMOS circuits/inverters. I_{ON} and I_{OFF} currents improve by changing the gate cylinder diameter. It was shown that this can be used for the P-type device in the CMOS circuit structure to improve its

ON and OFF current. Applying these optimisations to the P-type device in the CMOS stage shows that significant improvements in rise time and over/under shoot are achieved using asymmetric CMOS circuits.

ACKNOWLEDGEMENT

This research is financially supported by EPSRC grant number EP/E023150/1.

REFERENCES

1. K. Fobelets, P.W. Ding., and J. E. Velazquez-Perez, *A novel 3D Gate Field Effect Transistor Screen-Grid FET Device-Concept and Modelling.* Solid State Electron, 2007. **51 (5)**: p. 749-759.
2. P. W. Ding, K.Fobelets., and J.E. Velazquez-Perez, *Analog Performance of Screen Grid Field Effect Transistor (SGrFET).* WOFE, Cozumel (Mexico) 2007.
3. *Taurus-Medici User Guide (Version W-2004.09), S.I., Mountain View, CA.*
4. B. Andreev, E. L. Titlebaum and E. G. Friedman., *Sizing CMOS Inverter with Miller Effect and Threshold voltage Variations.* World Science, 2006. **15**(5): p. 437-454.
5. Y. Shadrokh, K.Fobelets., and J. E. Velazquez-Perez, *Single Device Logic using 3D Gating of Screen Grid Feild Effect Transistor Logic.* Semiconductor Conference, 2007. CAS 2007. International, October 2007. **1**: p. 45-48.